THE PICKERING MASTERS

THE WORKS OF
CHARLES DARWIN

Volume 18. *Movements and habits of climbing plants*

THE WORKS OF
CHARLES DARWIN

EDITED BY
PAUL H. BARRETT & R. B. FREEMAN
ADVISOR: PETER GAUTREY

VOLUME
18

THE MOVEMENTS AND HABITS
OF CLIMBING PLANTS

Routledge
Taylor & Francis Group

LONDON

First published 1992 by Pickering & Chatto (Publishers) Limited

Published 2016 by Routledge
2 Park Square, Milton Park, Abingdon, Oxon OX14 4RN

Routledge is an imprint of the Taylor & Francis Group, an informa business

Darwin, Charles, *1809–1882*
  The works of Charles Darwin.
  Vols. 11–20
  1. Organisms. Evolution – Early works
  I. Title II. Barrett, Paul H. (Paul
  Howard), *1917–1987* III. Freeman, R. B.
  (Richard Broke), *1915–1986* IV. Gautrey, Peter
  575
  ISBN 13: 978-1-85196-308-9 (hbk)

INTRODUCTION TO VOLUME EIGHTEEN

Movements and Habits of Climbing Plants. [Second edition] 1882. Revised, third thousand. Freeman 839.

Darwin's new hothouses at Down House were built early in 1862. For the next three years, when his health was being particularly difficult, he was able to observe nutation in climbing plants both in these houses and in his study. In 1865, he was able to write up his findings in the *Journal of the Proceedings of the Linnean Society of London* Vol. 9, pages 1–128 with thirteen text woodcuts. At the same time the publishers for the Society, Longmans and Williams & Norgate, produced the article as a paperback pamphlet; this is the first edition of the book. Ten years later, after some further work, he was able to offer a second edition to John Murray which was only slightly longer than the first. The text printed here is the third thousand which has an appendix to the preface; the figures are the same as those of the first edition.

This little book gives a marvellous impression of Darwin as an experimental scientist, practising a great deal of watching, a little touching, and a lot of measurement. It also contains much less than some of his works of the observations of others. At one point he describes how 'the plant tried but failed to entwine a pole 5 inches in diameter, where previously it succeeded in twisting itself round a 4 inch pole'. Such statements show that he treated all living beings as adaptive organisms, their morphological structures serving the 'purpose of contributing to survival'.

Darwin has been adversely criticized for being too teleological in his approach, but in reality teleological explanations, if cautiously employed, and if they serve the function of stimulating creative investigations, can be useful. Moving from a teleological position to a more mechanistic stance was a convenient device for drawing out function, form, and cause and effect. He often dealt with living things as if they had 'goal-orientated' actions, and by so doing he could move more easily into mechanistic speculations about physical and biological laws at work.

*The Movements and Habits of
Climbing Plants*

THE

MOVEMENTS AND HABITS

OF

CLIMBING PLANTS.

By CHARLES DARWIN, M.A., F.R.S.,

ETC.

SECOND EDITION, REVISED.

WITH ILLUSTRATIONS.

LONDON:
JOHN MURRAY, ALBEMARLE STREET.
1875.

PREFACE

This Essay first appeared in the ninth volume of the *Journal of the Linnean Society*, published in 1865. It is here reproduced in a corrected and, I hope, clearer form, with some additional facts. The illustrations were drawn by my son, George Darwin. Fritz Müller, after the publication of my paper, sent to the Linnean Society (*Journal*, vol. ix, p. 344) some interesting observations on the climbing plants of South Brazil, to which I shall frequently refer. Recently two important memoirs, chiefly on the difference in growth between the upper and lower sides of tendrils, and on the mechanism of the movements of twining plants, by Dr Hugo de Vries, have appeared in the *Arbeiten des Botanischen Instituts in Würzburg*, Heft. iii, 1873. These memoirs ought to be carefully studied by everyone interested in the subject, as I can here give only references to the more important points. This excellent observer, as / well as Professor Sachs,[1] attributes all the movements of tendrils to rapid growth along one side; but, from reasons assigned towards the close of my fourth chapter, I cannot persuade myself that this holds good with respect to those due to a touch. In order that the reader may know what points have interested me most, I may call his attention to certain tendril-bearing plants; for instance, *Bignonia capreolata*, Cobaea, Echinocystis, and Hanburya, which display as beautiful adaptations as can be found in any part of the kingdom of nature. It is, also, an interesting fact that intermediate states between organs fitted for widely different functions, may be observed on the same individual plant of *Corydalis claviculata* and the common vine; and these cases illustrate in a striking manner the principle of the gradual evolution of species. /

[1] An English translation of the *Lehrbuch der Botanik* by Professor Sachs, has recently (1875), appeared under the title of *Text-Book of Botany*, and this is a great boon to all lovers of natural science in England.

CONTENTS

CHAPTER I

Twining plants

Introductory remarks – Description of the twining of the hop – Torsion of the stems – Nature of the revolving movement, and manner of ascent – Stems not irritable – Rate of revolution in various plants – Thickness of the support round which plants can twine – Species which revolve in an anomalous manner

CHAPTER II

Leaf-climbers

Plants which climb by the aid of spontaneously revolving and sensitive petioles – Clematis – Tropaeolum – Maurandia, flower-peduncles moving spontaneously and sensitive to a touch – Rhodochiton – Lophospermum, internodes sensitive – Solanum, thickening of the clasped petioles – Fumaria – Adlumia – Plants which climb by the aid of their produced midribs – Gloriosa – Flagellaria – Nepenthes – Summary on leaf-climbers

CHAPTER III

Tendril-bearers

Nature of tendrils – Bignoniaceae, various species of, and their different modes of climbing – Tendrils which avoid the light, and creep into crevices – Development of adhesive discs – Excellent adaptations for seizing different kinds of supports – Polemoniaceae –

CHAPTER IV

Tendril-bearers continued

CHAPTER V

Hook and root-climbers – Concluding remarks

CHAPTER I

TWINING PLANTS

Introductory remarks – Description of the twining of the hop – Torsion of the stems – Nature of the revolving movement, and manner of ascent – Stems not irritable – Rate of revolution in various plants – Thickness of the support round which plants can twine – Species which revolve in an anomalous manner

I was led to this subject by an interesting, but short paper by Professor Asa Gray on the movements of the tendrils of some cucurbitaceous plants.[1] My observations were more than half completed before I learnt that the surprising phenomenon of the spontaneous revolutions of the stems and tendrils of climbing plants had been long ago observed by Palm and by Hugo von Mohl,[2] and had subsequently been the subject of two memoirs by Dutrochet.[3] Nevertheless, / I believe that my observations, founded on the examination of above a hundred widely distinct living species, contain sufficient novelty to justify me in publishing them.

Climbing plants may be divided into four classes. First, those which twine spirally round a support, and are not aided by any other movement. Secondly, those endowed with irritable organs, which when they touch any object clasp it; such organs consisting of modified leaves, branches, or flower-peduncles. But these two classes sometimes graduate to a certain extent into one another. Plants of the third class ascend merely by the aid of hooks; and those of the fourth by rootlets; but as in neither class do the plants exhibit any special movements, they present little interest, and generally when I speak of climbing plants I refer to the two first great classes.

[1] *Proc. Amer. Acad. of Arts and Sciences*, vol. iv, August, 12, 1858, p. 98.

[2] Ludwig H. Palm, *Ueber das Winden der Pflanzen*; Hugo von Mohl, *Ueber den Bau und das Winden der Ranken und Schlingpflanzen*, 1827. Palm's Treatise was published only a few weeks before Mohl's. See also *The Vegetable Cell* (translated by Henfrey), by H. von Mohl, p. 147 to end.

[3] 'Des Mouvements révolutifs / spontanés', *Comptes Rendus*, vol. xvii, 1843, p. 989; 'Recherches sur la Volubilité des Tiges', vol. xix, 1844, p. 295.

TWINING PLANTS

This is the largest subdivision, and is apparently the primordial and simplest condition of the class. My observations will be best given by taking a few special cases. When the shoot of a hop (*Humulus lupulus*) rises from the ground, the two or three first-formed joints or internodes are straight and remain stationary; but the next-formed, whilst very young, / may be seen to bend to one side and to travel slowly round towards all points of the compass, moving, like the hands of a watch, with the sun. The movement very soon acquires its full ordinary velocity. From seven observations made during August on shoots proceeding from a plant which had been cut down, and on another plant during April, the average rate during hot weather and during the day is 2 hrs 8 m for each revolution; and none of the revolutions varied much from this rate. The revolving movement continues as long as the plant continues to grow; but each separate internode, as it becomes old, ceases to move.

To ascertain more precisely what amount of movement each internode underwent, I kept a potted plant, during the night and day, in a well-warmed room to which I was confined by illness. A long shoot projected beyond the upper end of the supporting stick, and was steadily revolving. I then took a longer stick and tied up the shoot, so that only a very young internode, 1¾ of an inch in length, was left free. This was so nearly upright that its revolution could not be easily observed; but it certainly moved, and the side of the internode which was at one time convex became concave, which, as we shall hereafter see, is a sure sign of the revolving movement. I will assume that it made at least one revolution during the first twenty-four hours. Early the next morning its position was marked, and it made a second revolution in 9 hrs; during the latter part of this revolution it moved much quicker, and the third circle was performed in the evening in a little over / 3 hrs. As on the succeeding morning I found that the shoot revolved in 2 hrs 45 m, it must have made during the night four revolutions, each at the average rate of a little over 3 hrs. I should add that the temperature of the room varied only a little. The shoot had now grown 3½ inches in length, and carried at its extremity a young internode 1 inch in length, which showed slight changes in its curvature. The next or ninth revolution was effected in 2 hrs 30 m. From this time forward, the revolutions were easily observed. The

2

thirty-sixth revolution was performed at the usual rate; so was the last or thirty-seventh, but it was not completed; for the internode suddenly became upright, and after moving to the centre, remained motionless. I tied a weight to its upper end, so as to bow it slightly and thus detect any movement; but there was none. Some time before the last revolution was half performed, the lower part of the internode ceased to move.

A few more remarks will complete all that need be said about this internode. It moved during five days; but the more rapid movements, after the performance of the third revolution, lasted during three days and twenty hours. The regular revolutions, from the ninth to thirty-sixth inclusive, were effected at the average rate of 2 hrs 31 m; but the weather was cold, and this affected the temperature of the room, especially during the night, and consequently retarded the rate of movement a little. There was only one irregular movement, which consisted in the stem rapidly making, after an unusually slow revolution, only the / segment of a circle. After the seventeenth revolution the internode had grown from 1¾ to 6 inches in length, and carried an internode 1⅞ inch long, which was just perceptibly moving; and this carried a very minute ultimate internode. After the twenty-first revolution, the penultimate internode was 2½ inches long, and probably revolved in a period of about three hours. At the twenty-seventh revolution the lower and still moving internode was 8⅜, the penultimate 3½, and the ultimate 2½ inches in length; and the inclination of the whole shoot was such, that a circle 19 inches in diameter was swept by it. When the movement ceased, the lower internode was 9 inches, and the penultimate 6 inches in length; so that, from the twenty-seventh to thirty-seventh revolutions inclusive, three internodes were at the same time revolving.

The lower internode, when it ceased revolving, became upright and rigid; but as the whole shoot was left to grow unsupported, it became after a time bent into a nearly horizontal position, the uppermost and growing internodes still revolving at the extremity, but of course no longer round the old central point of the supporting stick. From the changed position of the centre of gravity of the extremity, as it revolved, a slight and slow swaying movement was given to the long horizontally projecting shoot; and this movement I at first thought was a spontaneous one. As the shoot grew, it hung down more and more, whilst the growing and revolving extremity turned itself up more and more.

With the hop we have seen that three internodes / were at the same time revolving; and this was the case with most of the plants observed by me. With all, if in full health, two internodes revolved; so that by the time the lower one ceased to revolve, the one above was in full action, with a terminal internode just commencing to move. With *Hota carnosa*, on the other hand, a depending shoot, without any developed leaves, 32 inches in length, and consisting of seven internodes (a minute terminal one, an inch in length, being counted), continually, but slowly, swayed from side to side in a semicircular course, with the extreme internodes making complete revolutions. This swaying movement was certainly due to the movement of the lower internodes, which, however, had not force sufficient to swing the whole shoot round the central supporting stick. The case of another asclepiadaceous plant, viz., *Ceropegia Gardnerii*, is worth briefly giving. I allowed the top to grow out almost horizontally to the length of 31 inches; this now consisted of three long internodes, terminated by two short ones. The whole revolved in a course opposed to the sun (the reverse of that of the hop), at rates between 5 hrs 15 m and 6 hrs 45 m for each revolution. The extreme tip thus made a circle of above 5 feet (or 62 inches) in diameter and 16 feet in circumference, travelling at the rate of 32 or 33 inches per hour. The weather being hot, the plant was allowed to stand on my study table; and it was an interesting spectacle to watch the long shoot sweeping this grand circle, night and day, in search of some object round which to twine. /

If we take hold of a growing sapling, we can of course bend it to all sides in succession, so as to make the tip describe a circle, like that performed by the summit of a spontaneously revolving plant. By this movement the sapling is not in the least twisted round its own axis. I mention this because if a black point be painted on the bark, on the side which is uppermost when the sapling is bent towards the holder's body, as the circle is described, the black point gradually turns round and sinks to the lower side, and comes up again when the circle is completed; and this gives the false appearance of twisting, which, in the case of spontaneously revolving plants, deceived me for a time. The appearance is the more deceitful because the axes of nearly all twining-plants are really twisted; and they are twisted in the same direction with the spontaneous revolving movement. To give an instance, the internode of the hop of which the history has been recorded, was at first, as could be seen by the ridges on its surface, not in the least twisted; but when, after the 37th revolution, it had grown 9 inches long, and its revolving movement had ceased, it had become twisted three times round its own axis, in the line of the course of the sun; on the other hand, the common convolvulus, which revolves in an opposite course to the hop, becomes twisted in an opposite direction.

Hence it is not surprising that Hugo von Mohl (p. 105, 108, etc.) thought that the twisting of the axis caused the revolving movement; but it is not / possible that the twisting of the axis of the hop three times should have caused thirty-seven revolutions. Moreover, the revolving movement commenced in the young internode before any twisting of its axis could be detected. The internodes of a young Siphomeris and Lecontea revolved during several days, but became twisted only once round their own axes. The best evidence, however, that the twisting does not cause the revolving movement is afforded by many leaf-climbing and tendril-bearing plants (as *Pisum sativum, Echinocystis lobata, Bignonia capreolata, Eccremocarpus scaber*, and with the leaf-climbers, *Solanum jasminoides* and various species of *Clematis*), of which the internodes are not twisted, but which, as we shall hereafter see, regularly perform revolving movements like those of true twining plants. Moreover, according to Palm (pp. 30, 95) and Mohl (p. 149), and Léon,[4] internodes may occasionally, and even not very rarely, be found which are twisted in an opposite direction to the other internodes on the same plant, and to the course of their revolutions; and this, according to Léon (p. 356), is the case with all the internodes of a certain variety of *Phaseolus multiflorus*. Internodes which have become twisted round their own axes, if they have not ceased to revolve, are still capable of twining round a support, as I have several times observed.

Mohl has remarked (p. 111) that when a stem twines round a smooth cylindrical stick, it does not become / twisted.[5] Accordingly I allowed kidney-beans to run up stretched string, and up smooth rods of iron and glass, one-third of an inch in diameter, and they became twisted only in that degree which follows as a mechanical necessity from the spiral winding. The stems, on the other hand, which had ascended ordinary rough sticks were all more or less and generally much

---

[4] *Bull. Bot. Soc. de France*, vol. v, 1858, p. 356.

[5] This whole subject has been ably discussed and explained by H. de Vries, *Arbeiten des Bot. Instituts in Würzburg*, vol. iii, pp. 331, 336. See also Sachs (*Text-Book of Botany*, English translation, 1875, p. 770), who concludes 'that torsion is the result of growth continuing in the outer layers after it has ceased or begun to cease in the inner layers'.

twisted. The influence of the roughness of the support in causing axial twisting was well seen in the stems which had twined up the glass rods; for these rods were fixed into split sticks below, and were secured above to cross sticks, and the stems in passing these places became much twisted. As soon as the stems which had ascended the iron rods reached the summit and became free, they also became twisted; and this apparently occurred more quickly during windy than during calm weather. Several other facts could be given, showing that the axial twisting stands in some relation to inequalities in the support, and likewise to the shoot revolving freely without any support. Many plants, which are not twiners, become in some degree twisted round their own axes;[6] but this occurs so much more / generally and strongly with twining plants than with other plants, that there must be some connection between the capacity for twining and axial twisting. The stem probably gains rigidity by being twisted (on the same principle that a much twisted rope is stiffer than a slackly twisted one), and is thus indirectly benefited so as to be enabled to pass over inequalities in its spiral ascent, and to carry its own weight when allowed to revolve freely.[7]

I have alluded to the twisting which necessarily follows on mechanical principles from the spiral ascent of a stem, namely, one twist for each spire completed. This was well shown by painting straight lines on living stems, and then allowing them to twine; but, as I shall have to recur to this subject under tendrils, it may be here passed over.

The revolving movement of a twining plant has been compared with that of the tip of a sapling, moved round and round by the hand held some way down the stem; but there is one important difference. The upper part of the sapling when thus moved / remains straight; but with twining plants every part of the revolving shoot has its own separate and independent movement. This is easily proved; for when the lower

[6] Professor Asa Gray has remarked to me, in a letter, that in *Thuja occidentalis* the twisting of the bark is very conspicuous. The twist is generally to the right of the observer; but, in noticing about a hundred trunks, four or / five were observed to be twisted in an opposite direction. The Spanish chestnut is often much twisted: there is an interesting article on this subject in the *Scottish Farmer*, 1865, p. 833.

[7] It is well known that the stems of many plants occasionally become spirally twisted in a monstrous manner; and after my paper was read before the Linnean Society, Dr Maxwell Masters remarked to me in a letter that 'some of these cases, if not all, are dependent upon some obstacle or resistance to their upward growth'. This conclusion agrees with what I have said about the twisting of stems, which have twined round rugged supports; but does not preclude the twisting being of service to the plant by giving greater rigidity to the stem.

half or two-thirds of a long revolving shoot is tied to a stick, the upper free part continues steadily revolving. Even if the whole shoot, except an inch or two of the extremity, be tied up, this part, as I have seen in the case of the hop, *Ceropegia*, convolvulus, etc., goes on revolving, but much more slowly; for the internodes, until they have grown to some little length, always move slowly. If we look to the one, two, or several internodes of a revolving shoot, they will be all seen to be more or less bowed, either during the whole or during a large part of each revolution. Now if a coloured streak be painted (this was done with a large number of twining plants) along, we will say, the convex surface, the streak will after a time (depending on the rate of revolution) be found to be running laterally along one side of the bow, then along the concave side, then laterally on the opposite side, and, lastly, again on the originally convex surface. This clearly proves that during the revolving movement the internodes become bowed in every direction. The movement is, in fact, a continuous self-bowing of the whole shoot, successively directed to all points of the compass; and has been well designated by Sachs as a revolving nutation.

As this movement is rather difficult to understand, it will be well to give an illustration. Take a sapling and bend it to the south, and paint a black line on the / convex surface; let the sapling spring up and bend it to the east, and the black line will be seen to run along the lateral face fronting the north; bend it to the north, the black line will be on the concave surface; bend it to the west, the line will again be on the lateral face; and when again bent to the south, the line will be on the original convex surface. Now, instead of bending the sapling, let us suppose that the cells along its northern surface from the base to the tip were to grow much more rapidly than on the three other sides, the whole shoot would then necessarily be bowed to the south; and let the longitudinal growing surface creep round the shoot, deserting by slow degrees the northern side and encroaching on the western side, and so round by the south, by the east, again to the north. In this case the shoot would remain always bowed with the painted line appearing on the several above specified surfaces, and with the point of the shoot successively directed to each point of the compass. In fact, we should have the exact kind of movement performed by the revolving shoots of twining plants.[8]

[8] The view that the revolving movement or nutation of the stems of twining plants is due to growth is that advanced by Sachs and H. de Vries; and the truth of this view is proved by their excellent observations.

7

It must not be supposed that the revolving movement is as regular as that given in the above illustration; in very many cases the tip describes an ellipse, even a very narrow ellipse. To recur once again to / our illustration, if we suppose only the northern and southern surfaces of the sapling alternately to grow rapidly, the summit would describe a simple arc; if the growth first travelled a very little to the western face, and during the return a very little to the eastern face, a narrow ellipse would be described; and the sapling would be straight as it passed to and fro through the intermediate space; and a complete straightening of the shoot may often be observed in revolving plants. The movement is frequently such that three of the sides of the shoot seem to be growing in due order more rapidly than the remaining side; so that a semicircle instead of a circle is described, the shoot becoming straight and upright during half of its course.

When a revolving shoot consists of several internodes, the lower ones bend together at the same rate, but one or two of the terminal ones bend at a slower rate; hence, though at times all the internodes are in the same direction, at other times the shoot is rendered slightly serpentine. The rate of revolution of the whole shoot, if judged by the movement of the extreme tip, is thus at times accelerated or retarded. One other point must be noticed. Authors have observed that the end of the shoot in many twining plants is completely hooked; this is very general, for instance, with the Asclepiadaceae. The hooked tip, in all the cases observed by me, viz. in *Ceropegia, Sphaerostema, Clerodendron, Wistaria, Stephania, Akebia,* and *Siphomeris,* has exactly the same kind of movement as the / other internodes; for a line painted on the convex surface first becomes lateral and then concave; but, owing to the youth of these terminal internodes, the reversal of the hook is a slower process than that of the revolving movement.[9] This strongly marked tendency in the young, terminal and flexible internodes, to bend in a greater degree or more abruptly than the other internodes, is of service to the plant; for not only does the hook thus formed sometimes serve to catch a support, but (and this seems to be much more important) it causes the extremity of the shoot to embrace the support much more closely than it could otherwise have done, and thus aids in preventing the stem from being blown away during windy weather, as

[9] The mechanism by which the end of the shoot remains hooked appears to be a difficult and complex problem, discussed by Dr H. de Vries (ibid., p. 337): he concludes that 'it depends on the relation between the rapidity of torsion and the rapidity of nutation'.

I have many times observed. In *Lonicera brachypoda* the hook only straightens itself periodically, and never becomes reversed. I will not assert that the tips of all twining plants when hooked, either reverse themselves or become periodically straight, in the manner just described; for the hooked form may in some cases be permanent, and be due to the manner of growth of the species, as with the tips of the shoots of the common vine, and more plainly with those of *Cissus discolor* – plants which are not spiral twiners.

The first purpose of the spontaneous revolving movement, or, more strictly speaking, of the continuous / bowing movement directed successively to all points of the compass, is, as Mohl has remarked, to favour the shoot finding a support. This is admirably effected by the revolutions carried on night and day, a wider and wider circle being swept as the shoot increases in length. This movement likewise explains how the plants twine; for when a revolving shoot meets with a support, its motion is necessarily arrested at the point of contact, but the free projecting part goes on revolving. As this continues, higher and higher points are brought into contact with the support and are arrested; and so onwards to the extremity; and thus the shoot winds round its support. When the shoot follows the sun in its revolving course, it winds round the support from right to left, the support being supposed to stand in front of the beholder; when the shoot revolves in an opposite direction, the line of winding is reversed. As each internode loses from age its power of revolving, it likewise loses its power of spirally twining. If a man swings a rope round his head, and the end hits a stick, it will coil round the stick according to the direction of the swinging movement; so it is with a twining plant, a line of growth travelling round the free part of the shoot causing it to bend towards the opposite side, and this replaces the momentum of the free end of the rope.

All the authors, except Palm and Mohl, who have discussed the spiral twining of plants, maintain that such plants have a natural tendency to grow spirally. Mohl believes (p. 112) that twining stems have / a dull kind of irritability, so that they bend towards any object which they touch; but this is denied by Palm. Even before reading Mohl's interesting treatise, this view seemed to me so probable that I tested it in every way that I could, but always with a negative result. I rubbed many shoots much harder than is necessary to excite movement in any tendril or in the footstalk of any leaf climber, but without any effect. I then tied a light forked twig to a shoot of a hop, a

9

Ceropegia, Sphaerostema, and *Adhatoda,* so that the fork pressed on one side alone of the shoot and revolved with it; I purposely selected some very slow revolvers, as it seemed most likely that these would profit most from possessing irritability; but in no case was any effect produced.[10] Moreover, when a shoot winds round a support, the winding movement is always slower, as we shall immediately see, than whilst it revolves freely and touches nothing. Hence I conclude that twining stems are not irritable; and indeed it is not probable that they should be so, as nature always economizes her means, and irritability would have been superfluous. Nevertheless I do not wish to assert that they are never irritable; for the growing axis of the leaf-climbing, but not spirally twining, *Lophospermum scandens* is, certainly irritable; but this case gives me confidence that ordinary twiners / do not possess any such quality, for directly after putting a stick to the *Lophospermum,* I saw that it behaved differently from a true twiner or any other leaf-climber.[11]

The belief that twiners have a natural tendency to grow spirally, probably arose from their assuming a spiral form when wound round a support, and from the extremity, even whilst remaining free, sometimes assuming this form. The free internodes of vigorously growing plants, when they cease to revolve, become straight, and show no tendency to be spiral; but when a shoot has nearly ceased to grow, or when the plant is unhealthy, the extremity does occasionally become spiral. I have seen this in a remarkable manner with the ends of the shoots of the *Stauntonia* and of the allied *Akebia,* which became wound up into a close spire, just like a tendril; and this was apt to occur after some small, ill-formed leaves had perished. The explanation, I believe, is, that in such cases the lower parts of the terminal internodes very gradually and successively lose their power of movement, whilst the portions just above move onwards and in their turn become motionless; and this ends in forming an irregular spire.

When a revolving shoot strikes a stick, it winds round it rather more slowly than it revolves. For instance, a shoot of the *Ceropegia,* revolved in 6 hrs, / but took 9 hrs 30 m to make one complete spire round a stick; *Aristolochia gigas* revolved in about 5 hrs, but took 9 hrs 15 m to

---

[10] Dr H. de Vries also has shown (ibid., p. 321 and 325) by a better method than that employed by me, that the stems of twining plants are not irritable, and that the cause of their winding up a support is exactly what I have described.

[11] Dr H. de Vries states (ibid., p. 322) that the stem of Cuscuta is irritable like a tendril.

complete its spire. This, I presume, is due to the continued disturbance of the impelling force by the arrestment of the movement at successive points; and we shall hereafter see that even shaking a plant retards the revolving movement. The terminal internodes of a long, much-inclined, revolving shoot of the *Ceropegia*, after they had wound round a stick, always slipped up it, so as to render the spire more open than it was at first; and this was probably in part due to the force which caused the revolutions, being now almost freed from the constraint of gravity and allowed to act freely. With the *Wistaria*, on the other hand, a long horizontal shoot wound itself at first into a very close spire, which remained unchanged; but subsequently, as the shoot twined spirally up its support, it made a much more open spire. With all the many plants which were allowed freely to ascend a support, the terminal internodes made at first a close spire; and this, during windy weather, served to keep the shoots in close contact with their support; but as the penultimate internodes grew in length, they pushed themselves up for a considerable space (ascertained by coloured marks on the shoot and on the support) round the stick, and the spire became more open.[12]

It follows from this latter fact that the position / occupied by each leaf with respect to the support, depends on the growth of the internodes after they have become spirally wound round it. I mention this on account of an observation by Palm (p. 34), who states that the opposite leaves of the hop always stand in a row, exactly over one another, on the same side of the supporting stick, whatever its thickness may be. My sons visited a hop-field for me, and reported that though they generally found the points of insertion of the leaves standing over each other for a space of two or three feet in height, yet this never occurred up the whole length of the pole; the points of insertion forming, as might have been expected, an irregular spire. Any irregularity in the pole entirely destroyed the regularity of position of the leaves. From casual inspection, it appeared to me that the opposite leaves of *Thunbergia alata* were arranged in lines up the sticks round which they had twined; accordingly, I raised a dozen plants, and gave them sticks of various thicknesses, as well as string, to twine round; and in this case one alone out of the dozen had its leaves arranged in a perpendicular line: I conclude, therefore, Palm's statement is not quite accurate.

[12] See Dr H. de Vries (ibid., p. 324) on this subject.

The leaves of different twining plants are arranged on the stem (before it has twined) alternately, or oppositely, or in a spire. In the latter case the line of insertion of the leaves and the course of the revolutions coincide. This fact has been well shown by Dutrochet,[11] / who found different individuals of *Solanum dulcamara* twining in opposite directions, and these had their leaves in each case spirally arranged in the same direction. A dense whorl of many leaves would apparently be incommodious for a twining plant, and some authors assert that none have their leaves thus arranged; but a twining *Siphomeris* has whorls of three leaves.

If a stick which has arrested a revolving shoot, but has not as yet been encircled, be suddenly taken away, the shoot generally springs forward, showing that it was pressing with some force against the stick. After a shoot has wound round a stick, if this be withdrawn, it retains for a time its spiral form; it then straightens itself, and again commences to revolve. The long, much-inclined shoot of the *Ceropegia* previously alluded to offered some curious peculiarities. The lower and older internodes, which continued to revolve, were incapable, on repeated trials, of twining round a thin stick; showing that, although the power of movement was retained, this was not sufficient to enable the plant to twine. I then moved the stick to a greater distance, so that it was struck by a point 2½ inches from the extremity of the penultimate internode; and it was then neatly encircled by this part of the penultimate and by the ultimate internode. After leaving the spirally wound shoot for eleven hours, I quietly withdraw the stick, and in the course of the day the curled portion straightened itself and recommenced revolving; but the lower and not curled portion of the penultimate internode did / not move, a sort of hinge separating the moving and the motionless part of the same internode. After a few days, however, I found that this lower part had likewise recovered its revolving power. These several facts show that the power of movement is not immediately lost in the arrested portion of a revolving shoot; and that after being temporarily lost it can be recovered. When a shoot has remained for a considerable time round a support, it permanently retains its spiral form even when the support is removed.

When a tall stick was placed so as to arrest the lower and rigid internodes of the *Ceropegia*, at the distance at first of 15 and then of 21

[13] *Comptes Rendus*, 1844, vol. xix, p. 295, and *Annales des Sc. Nat. 3rd series, Bot.*, vol. ii, p. 163.

inches from the centre of revolution, the straight shoot slowly and gradually slid up the stick, so as to become more and more highly inclined, but did not pass over the summit. Then, after an interval sufficient to have allowed of a semi-revolution, the shoot suddenly bounded from the stick and fell over to the opposite side or point of the compass, and reassumed its previous slight inclination. It now recommenced revolving in its usual course, so that after a semi-revolution it again came into contact with the stick, again slid up it, and again bounded from it and fell over to the opposite side. This movement of the shoot had a very odd appearance, as if it were disgusted with its failure but was resolved to try again. We shall, I think, understand this movement by considering the former illustration of the sapling, in which the growing surface was supposed to creep round / from the northern by the western to the southern face; and thence back again by the eastern to the northern face, successively bowing the sapling in all directions. Now with the *Ceropegia*, the stick being placed to the south of the shoot and in contact with it, as soon as the circulatory growth reached the western surface, no effect would be produced, except that the shoot would be pressed firmly against the stick. But as soon as growth on the southern surface began, the shoot would be slowly dragged with a sliding movement up the stick; and then, as soon as the eastern growth commenced, the shoot would be drawn from the stick, and its weight coinciding with the effects of the changed surface of growth, would cause it suddenly to fall to the opposite side, reassuming its previous slight inclination; and the ordinary revolving movement would then go on as before. I have described this curious case with some care, because it first led me to understand the order in which, as I then thought, the surfaces contracted; but in which, as we now know from Sachs and H. de Vries, they grow for a time rapidly, thus causing the shoot to bow towards the opposite side.

The view just given further explains, as I believe, a fact observed by Mohl (p. 135), namely, that a revolving shoot, though it will twine round an object as thin as a thread, cannot do so round a thick support. I placed some long revolving shoots of a *Wistaria* close to a post between 5 and 6 inches in diameter, but, though aided by me in many ways, they could / not wind round it. This apparently was due to the flexure of the shoot, whilst winding round an object so gently curved as this post, not being sufficient to hold the shoot to its place when the growing surface crept round to the opposite surface of the shoot; so that it was withdrawn at each revolution from its support.

When a free shoot has grown far beyond its support, it sinks downwards from its weight, as already explained in the case of the hop, with the revolving extremity turned upwards. If the support be not lofty, the shoot falls to the ground, and resting there, the extremity rises up. Sometimes several shoots, when flexible, twine together into a cable, and thus support one another. Single thin depending shoots, such as those of the *Sollya Drummondii*, will turn abruptly backwards and wind up on themselves. The greater number of the depending shoots, however, of one twining plant, the *Hibbertia dentata*, showed but little tendency to turn upwards. In other cases, as with the *Cryptostegia grandiflora*, several internodes which were at first flexible and revolved, if they did not succeed in twining round a support, become quite rigid, and supporting themselves upright, carried on their summits the younger revolving internodes.

Here will be a convenient place to give a table showing the direction and rate of movement of several twining plants, with a few appended remarks. These plants are arranged according to Lindley's *Vegetable Kingdom* of 1853; and they have been selected from / all parts of the series so as to show that all kinds behave in a nearly uniform manner.[14]

The rate of revolution of various twining plants

ACOTYLEDONS

Lygodium scandens (Polypodiceae) moves against the sun.

	H	M
June 18, 1st circle was made in	6	0
18, 2nd	6	15 (late in evening)
19, 3rd	5	32 (very hot day)
19, 4th	5	0 (very hot day)
20, 5th	6	0

[14] I am much indebted to Dr Hooker for having sent me many plants from Kew; and to Mr Veitch, of the Royal Exotic Nursery, for having generously given me a collection of fine specimens of climbing plants. Professor Asa Gray, Professor Oliver, and Dr Hooker have afforded me, as on many previous occasions, much information and many references.

ACOTYLEDONS *continued*

Lygodium articulatum moves against the sun.

	H M
July 19, 1st circle was made in	16 30 (shoot very young)
20, 2nd	15 0
21, 3rd	8 0
22, 4th	10 30

MONOCOTYLEDONS

Ruscus androgynus (Liliaceae), placed in the hothouse, moves against the sun.

	H M
May 24, 1st circle was made in	6 14 (shoot very young)
25, 2nd	2 21
25, 3rd	3 37
25, 4th	3 22
26, 5th	2 50
27, 6th	3 52
27, 7th	4 11 /

Asparagus (unnamed species from Kew) (Liliaceae) moves against the sun, placed in hothouse.

	H M
Dec. 26, 1st circle was made in	5 0
27, 2nd	5 40

Tamus communis (Dioscoreaceae). A young shoot from a tuber in a pot placed in the greenhouse: follows the sun.

	H M
July 7, 1st circle was made in	3 10
7, 2nd	2 38
8, 3rd	3 5
8, 4th	2 56
8, 5th	2 30
8, 6th	2 30

MONOCOTYLEDONS *continued*

Lapagerea rosea (Philesiaceae), in greenhouse, follows the sun.

	H M
March 9, 1st circle was made in	26 15 (shoot young)
10, semicircle	8 15
11, 2nd circle	11 0
12, 3rd	15 30
13, 4th	14 15
16, 5th	8 40 when placed in the

hothouse; but the next day the shoot remained stationary.

Roxburghia viridiflora (Roxburghiaceae) moves against the sun; it completed a circle in about 24 hours.

DICOTYLEDONS

Humulus Lupulus (Urticaceae) follows the sun. The plant was kept in a room during warm weather.

	H M
April 9, 2 circles were made in	4 16
Aug. 13, 3rd circle was made in	2 0
14, 4th	2 20
14, 5th	2 16
14, 6th	2 2
14, 7th	2 0
14, 8th	2 4 /

With the hop a semicircle was performed, in travelling from the light, in 1 hr 33 m; in travelling to the light, in 1 hr 13 m; difference of rate, 20 m.

Akebia quinata (Lardizabalaceae), placed in hothouse, moves against the sun.

	H M
March 17, 1st circle was made in	4 0 (shoot young)
18, 2nd	1 40
18, 3rd	1 30
19, 4th	1 45

DICOTYLEDONS *continued*

Stauntonia latifolia (Lardizabalaceae), placed in hothouse, moves against the sun.

	H M
March 28, 1st circle was made in	3 30
29, 2nd	3 45

Sphaerostema marmoratum (Schizandraceae) follows the sun.

	H M
August 5, 1st circle was made in about	24 0
5, 2nd circle was made in	18 30

Stephania rotunda (Menispermaceae) moves against the sun.

	H M
May 27, 1st circle was made in	5 5
30, 2nd	7 6
June 2, 3rd	5 15
3, 4th	6 28

Thryallis brachystachys (Malpighiaceae) moves against the sun: one shoot made a circle in 12 hrs, and another in 10 hrs 30 m; but the next day, which was much colder, the first shoot took 10 hrs to perform only a semicircle.

Hibbertia dentata (Dilleniaceae), placed in the hothouse, followed the sun, and made (18 May) a circle in 7 hrs 20 m; on the 19th, reversed its course, and moved against the sun, and made a circle in 7 hrs; on the 20th, moved against the sun one-third of a circle, and then stood still; on the 26th, followed the / sun for two-thirds of a circle, and then returned to its starting-point, taking for this double course 11 hrs 46 m.

Sollya Drummondii (Pittosporaceae) moves against the sun; kept in greenhouse.

	H M
April 4, 1st circle was made in	4 25
5, 2nd	8 0 (very cold day)
6, 3rd	6 25
7, 4th	7 5

DICOTYLEDONS *continued*

Polygonum dumetorum (Polygonaceae). This case is taken from Dutrochet (p. 299), as I observed, no allied plant: follows the sun. Three shoots, cut off a plant, and placed in water, made circles in 3 hrs 10 m, 5 hrs 20 m, and 7 hrs 15 m.

Wistaria Chinensis (Leguminosae), in greenhouse, moves against the sun.

	H	M
May 13, 1st circle was made in	3	5
13, 2nd	3	20
16, 3rd	2	5
24, 4th	3	21
25, 5th	2	37
25, 6th	2	35

Phaseolus vulgaris (Leguminosae), in greenhouse, moves against the sun.

	H	M
May 1st circle was made in	2	0
2nd	1	55
3rd	1	55

Dipladenia urophylla (Apocynaceae) moves against the sun.

	H	M
April 18, 1st circle was made in	8	0
19, 2nd	9	15
30, 3rd	9	40

Dipladenia crassinoda moves against the sun.

	H	M
May 16, 1st circle was made in	9	5
July 20, 2nd	8	0
21, 3rd	8	5 /

DICOTYLEDONS *continued*

Ceropegia Gardnerii (Asclepiadaceae) moves against the sun.

		H M
Shoot very young, 2 inches in length	1st circle was performed in	7 55
Shoot still young	2nd	7 0
Long shoot	3rd	6 33
Long shoot	4th	5 15
Long shoot	5th	6 45

Stephanotis floribunda (Asclepiadaceae) moves against the sun and made a circle in 6 hrs 40 m, a second circle in about 9 hrs.

Hoya carnosa (Asclepiadaceae) made several circles in from 16 hrs to 22 hrs or 24 hrs.

Ipomoea purpurea (Convolvulaceae) moves against the sun. Plant placed in room with lateral light.

1st circle was made in 2 hrs 42 m	Semicircle, from the light in 1 hr 14 m, to the light 1 hr 28 m: difference 14 m.
2nd circle was made in 2 hrs 47 m	Semicircle, from the light in 1 hr 17 m, to the light 1 hr 30 m: difference 13 m.

Ipomoea jucunda (Convolvulaceae) moves against the sun, placed in my study, with windows facing the north-east. Weather hot.

1st circle was made in 5 hrs 30 m	Semicircle, from the light in 4 hrs 30 m, to the light 1 hr 0 m: difference 3 hrs 30 m.
2nd circle was made in 5 hrs 20 m. (Late in afternoon: circle completed at 6 hrs 40 m p.m.	Semicircle, from the light in 3 hrs 50 m, to the light 1 hr 30 m: difference 2 hrs 20 m.

We have here a remarkable instance of the power of light in retarding and hastening the revolving movement.

Convolvulus sepium (large-flowered cultivated variety) moves against

DICOTYLEDONS *continued*

the sun. Two circles, were made each in 1 hr 42 m: difference in semicircle from and to the light 14 m. /

Rivea tilioefolia (Convulvulaceae) moves against the sun; made four revolutions in 9 hrs; so that, on an average, each was performed in 2 hrs 15 m.

Plumbago rosea (Plumbaginaceae) follows the sun. The shoot did not begin to revolve until nearly a yard in height; it then made a fine circle in 10 hrs 45 m. During the next few days it continued to move, but irregularly. On 15 August the shoot followed, during a period of 10 hrs 40 m, a long and deeply zigzag course and then made a broad ellipse. The figure apparently represented three ellipses, each of which averaged 3 hrs 33 m for its completion.

Jasminum pauciflorum, Bentham (Jasminaceae), moves against the sun. A circle was made in 7 hrs 15 m, and a second rather more quickly.

Clerodendrum Thomsonii (Verbenaceae) follows the sun.

	H M
April 12, 1st circle was made in	5 45 (shoot very young)
14, 2nd	3 30
18, a semicircle	5 0 (directly after the plant was shaken on being moved)
19, 3rd circle	3 0
20, 4th	4 20

Tecoma jasminoides (Bignoniaceae) moves against the sun.

	H M
March 17, 1st circle was made in	6 30
19, 2nd	7 0
22, 3rd	8 30 (very cold day)
24, 4th	6 45

DICOTYLEDONS *continued*

Thunbergia alata (Acanthaceae) moves against sun.

	H M
April 14, 1st circle was made in	3 20
18, 2nd	2 50
18, 3rd	2 55
18, 4th	3 55 (late in afternoon) /

Adhadota cydonaefolia (Acanthaceae) follows the sun. A young shoot made a semicircle in 24 hrs; subsequently it made a circle in between 40 hrs and 48 hrs. Another shoot, however, made a circle in 26 hrs 30 m.

Mikania scandens (Compositae) moves against the sun.

	H M
March 14, 1st circle was made in	3 10
15, 2nd	3 0
16, 3rd	3 0
17, 4th	3 33
April 7, 5th	2 50
7, 6th	2 40 (this circle was made after a copious watering with cold water at 47° Fahr.)

Combretum argenteum (Combretaceae) moves against the sun. Kept in hothouse.

	H M
Jan. 24, 1st circle was made in	2 55 (early in morning, when the temperature of the house had fallen a little)
24, 2 circles each at an average of	2 20
25, 4th circle was made in	2 25

Combretum purpureum revolves not quite so quickly as *C. argenteum*.

21

DICOTYLEDONS *continued*

Loasa aurantiaca (Loasaceae). Revolution variable in their course: a plant which moved against the sun.

	H M
June 20, 1st circle was made in	2 37
20, 3rd	4 0
20, 2nd	2 13
21, 4th	2 35
22, 5th	3 26
23, 6th	3 5 /

Another plant which followed the sun in its revolutions.

	H M
July 11, 1st circle was made in	1 51 (very hot day)
11, 2nd	1 46 (very hot day)
11, 3rd	1 41 (very hot day)
11, 4th	1 48 (very hot day)
12, 5th	2 35

Scyphanthus elegans (Loasaceae) follows the sun.

	H M
June 13, 1st circle was made in	1 45
13, 2nd	1 17
14, 3rd	1 36
14, 4th	1 59
14, 5th	2 3

Siphomeris or *Lecontea* (unnamed sp.) (Cinchonaceae) follows the sun.

	H M
May 25, semicircle was made in	10 27 (shoot extremely young)
26, 1st circle was made in	10 15 (shoot still young)
30, 2nd	8 55
June 2, 3rd	8 11
6, 4th	6 8
8, 5th	7 20 (taken from the hot-
9, 6th	8 36 house, and placed in a room in my house)

DICOTYLEDONS *continued*

Manettia bicolor (Cinchonaceae), young plant, follows the sun.

	H	M
July 7, 1st circle was made in	6	18
8, 2nd	6	53
9, 3rd	6	30

Lonicera brachypoda (Caprifoliaceae) follows the sun, kept in a warm room in the house.

	H	M	
April 1st circle was made in	9	10	(about) /
2nd	12	20	(a distinct shoot, very young, on same plant)
3rd	7	30	
4th	8	0	(in this latter circle, the semicircle from the light took 5 hrs 23 m, and to the light 2 hrs 37 m: difference 2 hrs 46 m)

Aristolochia gigas (Aristolochiaceae) moves against the sun.

	H	M	
July 22, 1st circle was made in	8	0	(rather young shoot)
23, 2nd	7	15	
24, 3rd	5	0	(about)

In the foregoing table, which includes twining plants belonging to widely different orders, we see that the rate at which growth travels or circulates round the axis (on which the revolving movement depends), differs much. As long as a plant remains under the same conditions, the rate is often remarkably uniform, as with the hop, *Mikania*, *Phaseolus*, etc. The *Scyphanthus* made one revolution in 1 hr 17 m, and this is the quickest rate observed by me; but we shall hereafter see a tendril-bearing Passiflora revolving more rapidly. A shoot of the *Akebia quinata* made a revolution in 1 hr 30 m, and three revolutions at the average rate of 1 hr 38 m; a convolvulus made two revolutions at the average of 1 hr 42 m, and *Phaseolus vulgaris* three at the average of 1 hr 57 m. On the other hand, some plants take 24 hrs for a single revolution, and the *Adhadota* sometimes required 48 hrs; yet this latter

23

plant is an efficient twiner. / Species of the same genus move at different rates. The rate does not seem governed by the thickness of the shoots: those of the *Sollya* are as thin and flexible as string, but move more slowly than the thick and fleshy shoots of the *Ruscus*, which seem little fitted for movement of any kind. The shoots of the *Wistaria*, which become woody, move faster than those of the herbaceous *Ipomoea* or *Thunbergia*.

We know that the internodes, whilst still very young, do not acquire their proper rate of movement; hence the several shoots on the same plant may sometimes be seen revolving at different rates. The two or three, or even more, internodes which are first formed above the cotyledons, or above the rootstock of a perennial plant, do not move; they can support themselves, and nothing superfluous is granted.

A greater number of twiners revolve in a course opposed to that of the sun, or to the hands of a watch, than in the reversed course, and, consequently, the majority, as is well known, ascend their supports from left to right. Occasionally, though rarely, plants of the same order twine in opposite directions, of which Mohl (p. 125) gives a case in the Leguminosae, and we have in the table another in the Acanthaceae. I have seen no instance of two species of the same genus twining in opposite directions, and such cases must be rare; but Fritz Müller[15] states that although *Mikania / scandens* twines, as I have described, from left to right, another species in South Brazil twines in an opposite direction. It would have been an anomalous circumstance if no such cases had occurred, for different individuals of the same species, namely, of *Solanum dulcamara* (Dutrochet, vol. xix, p. 299), revolve and twine in two directions: this plant, however, is a most feeble twiner. *Loasa aurantiaca* (Léon, p. 351) offers a much more curious case: I raised seventeen plants: of these eight revolved in opposition to the sun and ascended from left to right; five followed the sun and ascended from right to left; and four revolved and twined first in one direction, and then reversed their course,[16] the petioles of the opposite leaves affording a *point d'appui* for the reversal of the spine. One of these four plants made seven spiral turns from right to left, and five turns from left to right. Another plant in the same family, the

---

[15] *Journal of the Linn. Soc.* (Bot.), vol. ix, p. 344. I shall have occasion often to quote this interesting paper, in which he corrects or confirms various statements made by me.
[16] I raised nine plants of the hybrid *Loasa Herbertii*, and six of these also reversed their spire in ascending a support.

Scyphanthus elegans, habitually twines in this same manner. I raised many plants of it, and the stems of all took one turn, or occasionally two or even three turns in one direction, and then, ascending for a short space straight, reversed their course and took one or two turns in an opposite direction. The reversal of the curvature occurred at any point in the stem, even in the middle of an internode. Had I not seen this case, I should have thought its occurrence / most improbable. It would be hardly possible with any plant which ascended above a few feet in height, or which lived in an exposed situation; for the stem could be pulled away easily from its support, with but little unwinding; nor could it have adhered at all, had not the internodes soon become moderately rigid. With leaf-climbers, as we shall soon see, analogous cases frequently occur; but these present no difficulty, as the stem is secured by the clasping petioles.

In the many other revolving and twining plants observed by me, I never but twice saw the movement reversed; once, and only for a short space, in *Ipomoea jucunda*; but frequently with *Hibbertia dentata*. This plant at first perplexed me much, for I continually observed its long and flexible shoots, evidently well fitted for twining, make a whole, or half, or quarter circle in one direction and then in an opposite direction; consequently, when I placed the shoots near thin or thick sticks, or perpendicularly stretched string, they seemed as if constantly trying to ascend, but always failed. I then surrounded the plant with a mass of branched twigs; the shoots ascended, and passed through them, but several came out laterally, and their depending extremities seldom turned upwards as is usual with twining plants. Finally, I surrounded a second plant with many thin upright sticks, and placed it near the first one with twigs; and now both had got what they liked, for they twined up the parallel sticks, sometimes winding round one and sometimes round several; and the shoots travelled / laterally from one to the other pot; but as the plants grew older, some of the shoots twined regularly up thin upright sticks. Though the revolving movement was sometimes in one direction and sometimes in the other, the twining was invariably from left to right;[7] so that the more potent or persistent movement of revolution must have been in opposition to

[7] In another genus, namely Davilla, belonging to the same family with Hibbertia, Fritz Müller says (ibid., p. 349) that 'the stem twines indifferently from left to right, or from right to left; and I once saw a shoot which ascended a tree about five inches in diameter, reverse its course in the same manner as so frequently occurs with Loasa'.

the course of the sun. It would appear that this *Hibbertia* is adapted both to ascend by twining, and to ramble laterally through the thick Australian scrub.

I have described the above case in some detail, because, as far as I have seen, it is rare to find any special adaptations with twining plants, in which respect they differ much from the more highly organized tendril-bearers. The *Solanum dulcamara*, as we shall presently see, can twine only round stems which are both thin and flexible. Most twining plants are adapted to ascend supports of moderate though of different thicknesses. Our English twiners, as far as I have seen, never twine round trees, excepting the honeysuckle (*Lonicera periclymenum*), which I have observed twining up a young beech-tree nearly 4½ inches in diameter. Mohl (p. 134) found that the *Phaseolus multiflorus* and *Ipomoea purpurea* could not, / when placed in a room with the light entering on one side, twine round sticks between 3 and 4 inches in diameter; for this interfered, in a manner presently to be explained, with the revolving movement. In the open air, however, the *Phaseolus* twined round a support of the above thickness, but failed in twining round one 9 inches in diameter. Nevertheless, some twiners of the warmer temperate regions can manage this latter degree of thickness; for I hear from Dr Hooker that at Kew the *Ruscus androgynus* has ascended a column 9 inches in diameter; and although a *Wistaria* grown by me in a small pot tried in vain for weeks to get round a post between 5 and 6 inches in thickness, yet at Kew a plant ascended a trunk above 6 inches in diameter. The tropical twiners, on the other hand, can ascend thicker trees; I hear from Dr Thomson and Dr Hooker that this is the case with the *Butea parviflora*, one of the Menispermaceae, and with some Dalbergias and other Leguminosae.[18] This power would be necessary for any species which had to ascend by twining the large trees of a tropical forest; otherwise they would hardly ever be able to reach the light. In our temperate countries it would be injurious to the twining plants which die down every year if / they were enabled to twine round trunks of trees, for they could not grow tall enough in a single season to reach the summit and gain the light.

By what means certain twining plants are adapted to ascend only

---

[18] Fritz Müller states (ibid., p. 349) that he saw on one occasion in the forests of South Brazil a trunk about five feet in circumference spirally ascended by a plant, apparently belonging to the Menispermaceae. He adds in his letter to me that most of the climbing plants which there ascend thick trees, are root-climbers; some being tendril-bearers.

thin stems, whilst others can twine round thicker ones, I do not know. It appeared to me probable that twining plants with very long revolving shoots would be able to ascend thick supports; accordingly I placed *Ceropegia Gardnerii* near a post 6 inches in diameter, but the shoots entirely failed to wind round it; their great length and power of movement merely aid them in finding a distant stem round which to twine. The *Sphaerostemma marmoratum* is a vigorous tropical twiner; and as it is a very slow revolver, I thought that this latter circumstance might help it in ascending a thick support; but though it was able to wind round a 6-inch post, it could do this only on the same level or plane, and did not form a spire and thus ascend.

As ferns differ so much in structure from phanerogamic plants, it may be worth while here to show that twining ferns do not differ in their habits from other twining plants. In *Lygodium articulatum* the two internodes of the stem (properly the rachis) which are first formed above the rootstock do not move; the third from the ground revolves, but at first very slowly. This species is a slow revolver: but *L. scandens* made five revolutions, each at the average rate of 5 hrs 45 m; and this represents fairly well the usual rate, taking quick and slow movers, among / phanerogamic plants. The rate was accelerated by increased temperature. At each stage of growth only the two upper internodes revolved. A line painted along the convex surface of a revolving internode becomes first lateral, then concave, then lateral and ultimately again convex. Neither the internodes nor the petioles are irritable when rubbed. The movement is in the usual direction, namely, in opposition to the course of the sun; and when the stem twines round a thin stick, it becomes twisted on its own axis in the same direction. After the young internodes have twined round a stick, their continued growth causes them to slip a little upwards. If the stick be soon removed, they straighten themselves, and recommence revolving. The extremities of the depending shoots turn upwards, and twine on themselves. In all these respects we have complete identity with twining phanerogamic plants; and the above enumeration may serve as a summary of the leading characteristics of all twining plants.

The power of revolving depends on the general health and vigour of the plant, as has been laboriously shown by Palm. But the movement of each separate internode is so independent of the others, that cutting off an upper one does not affect the revolutions of a lower one. When, however, Dutrochet cut off two whole shoots of the hop, and placed them in water, the movement was greatly retarded; for one revolved in

20 hrs and the other in 23 hrs, whereas they ought to have revolved in between 2 hrs and 2 hrs 30 m. / Shoots of the kidney-bean, cut off and placed in water, were similarly retarded, but in a less degree. I have repeatedly observed that carrying a plant from the greenhouse to my room, or from one part to another of the greenhouse, always stopped the movement for a time; hence I conclude that plants in a state of nature and growing in exposed situations, would not make their revolutions during very stormy weather. A decrease in temperature always caused a considerable retardation in the rate of revolution; but Dutrochet (vol. xvii, pp. 994, 996) has given such precise observations on this head with respect to the common pea that I need say nothing more. When twining plants are placed near a window in a room, the light in some cases has a remarkable power (as was likewise observed by Dutrochet, p. 998, with the pea) on the revolving movement, but this differs in degree with different plants; thus *Ipomoea jucunda* made a complete circle in 5 hrs 30 m; the semicircle from the light taking 4 hrs 30 m, and that towards the light only 1 hr. *Lonicera brachypoda* revolved, in a reversed direction to the *Ipomoeae*, in 8 hrs; the semicircle from the light taking 5 hrs 23 m, and that to the light only 2 hrs 37 m. From the rate of revolution in all the plants observed by me, being nearly the same during the night and the day, I infer that the action of the light is confined to retarding one semicircle and accelerating the other, so as not to modify greatly the rate of the whole revolution. This action of the light is remarkable, / when we reflect how little the leaves are developed on the young and thin revolving internodes. It is all the more remarkable, as botanists believe (Mohl, p. 119) that twining plants are but little sensitive to the action of light.

I will conclude my account of twining plants by giving a few miscellaneous and curious cases. With most twining plants all the branches, however many there may be, go on revolving together; but, according to Mohl (p. 4), only the lateral branches of *Tamus elephantipes* twine, and not the main stem. On the other hand, with a climbing species of asparagus, the leading shoot alone, and not the branches, revolved and twined; but it should be stated that the plant was not growing vigorously. My plants of *Combretum argenteum* and *C. purpureum* made numerous short healthy shoots; but they showed no signs of revolving, and I could not conceive how these plants could be climbers; but at last *C. argenteum* put forth from the lower part of one of its main branches a thin shoot, 5 or 6 feet in length, differing greatly in appearance from the previous shoots, owing to its leaves being little developed, and this shoot revolved vigorously and twined. So that this plant produces shoots of two kinds. With *Periploca Graeca* (Palm, p. 43) the uppermost shoots alone twine. *Polygonum convolvulus* twines only during the middle of the summer (Palm, p. 43, 94); and plants growing vigorously in the autumn show no inclination to climb. The majority of Asclepiadaceae are twiners; / but *Asclepias nigra* only 'in fertiliori solo incipit scandere subvolubili caule' (Willdenow, quoted and confirmed by Palm, p. 41). *Asclepias vincetoxicum* does not regularly twine, but occasionally does so (Palm, p. 42; Mohl, p. 112) when growing under certain conditions. So it is with two species of *Ceropegia*, as I hear from Professor Harvey, for these plants in their native dry South African home generally grow erect, from 6 inches to 2 feet in height – a very few taller specimens showing some inclination to curve; but when cultivated near Dublin, they regularly twined up sticks 5 or 6 feet in height. Most Convolvulaceae are excellent twiners; but in South Africa *Ipomoea argyraeoides* almost always grows erect and compact, from about 12 to 18 inches in height, one specimen alone in Professor Harvey's collection showing an evident disposition to twine. On the other hand, seedlings raised near Dublin twined up sticks above 8 feet in height. These facts are remarkable; for there can hardly be a doubt that in the dryer provinces of South Africa these plants have propagated themselves for thousands of generations in an erect condition; and yet they have retained during this whole period the innate power of spontaneously revolving and twining, whenever their shoots become elongated under proper conditions of life. Most of the species of *Phaseolus* are twiners; but certain varieties of the *P. multiflorus* produce (Léon, p. 681) two kinds of shoots, some upright and thick, and others thin and twining. I have seen striking instances of this curious / case of variability in 'Fulmer's dwarf forcing-bean', which occasionally produced a single long twining shoot.

Solanum dulcamara is one of the feeblest and poorest of twiners: it may often be seen growing as an upright bush, and when growing in the midst of a thicket merely scrambles up between the branches without twining; but when, according to Dutrochet (vol. xix, p. 299), it grows near a thin and flexible support, such as the stem of a nettle, it twines round it. I placed sticks round several plants, and vertically stretched strings close to others, and the strings alone were ascended by twining. The stem twines indifferently to the right or left. Some other species of Solanum, and of another genus, viz. *Habrothamnus*, belonging to the same family, are described in horticultural works as twining plants, but they seem to possess this faculty in a very feeble degree. We may suspect that the species of these two genera have as yet only partially acquired the habit of twining. On the other hand with *Tecoma radicans*, a member of a family abounding with twiners and tendril-bearers, but which climbs, like the ivy, by the aid of rootlets, we may suspect that a former habit of twining has been lost, for the stem exhibited slight irregular movements which could hardly be accounted for by changes in the action of the light. There is no difficulty in understanding how a spirally twining plant could graduate into a simple root-climber; for the young internodes of *Bignonia Tweedyana* and of *Hoya carnosa* revolve / and twine, but likewise emit rootlets which adhere to any fitting surface, so that the loss of twining would be no great disadvantage and in some respects an advantage to these species, as they would then ascend their supports in a more direct line.[19] /

[19] Fritz Müller has published some interesting facts and views on the structure of the wood of climbing plants in *Bot. Zeitung*, 1866, pp. 57, 65.

CHAPTER II

LEAF-CLIMBERS

Plants which climb by the aid of spontaneously revolving and sensitive petioles – Clematis – Tropaeolum – Maurandia, flower-peduncles moving spontaneously and sensitive to a touch – Rhodochiton – Lophospermum, internodes sensitive – Solanum, thickening of the clasped petioles – Fumaria – Adlumia – Plants which climb by the aid of their produced midribs – Gloriosa – Flagellaria – Nepenthes – Summary on leaf-climbers.

We now come to our second class of climbing plants, namely, those which ascend by the aid of irritable or sensitive organs. For convenience' sake the plants in this class have been grouped under two subdivisions, namely, leaf-climbers, or those which retain their leaves in a functional condition, and tendril-bearers. But these subdivisions graduate into each other, as we shall see under Corydalis and the Gloriosa lily.

It has long been observed that several plants climb by the aid of their leaves, either by their petioles (foot-stalks) or by their produced midribs; but beyond this simple fact they have not been described. Palm and Mohl class these plants with those which bear tendrils; but as a leaf is generally a defined object, the present classification, though artificial, has at least some advantages. Leaf-climbers are, moreover, intermediate in many respects between twiners and tendril-bearers. Eight species of clematis and seven of Tropaeolum were / observed, in order to see what amount of difference in the manner of climbing existed within the same genus; and the differences are considerable.

CLEMATIS

Clematis glandulosa. The thin upper internodes revolve, moving against the course of the sun, precisely like those of a true twiner, at an average rate, judging from three revolutions, of 3 hrs 48 m. The leading shoot immediately twined round a stick placed near it; but,

31

after making an open spire of only one turn and a half, it ascended for a short space straight, and then reversed its course and wound two turns in an opposite direction. This was rendered possible by the straight piece between the opposed spires having become rigid. The simple, broad, ovate leaves of this tropical species, with their short thick petioles, seem but ill-fitted for any movement; and whilst twining up a vertical stick, no use is made of them. Nevertheless, if the footstalk of a young leaf be rubbed with a thin twig a few times on any side, it will in the course of a few hours bend to that side; afterwards becoming straight again. The underside seemed to be the most sensitive; but the sensitiveness or irritability is slight compared to that which we shall meet with in some of the following species; thus, a loop of string, weighing 1·64 grain (106·2 mg) and hanging for some days on a young footstalk, produced a scarcely perceptible effect. A sketch is here given of two young leaves which had naturally caught hold of

Fig. 1   *Clematis glandulosa.*
With two young leaves clasping two twigs, with the clasping portions thickened

two thin branches. A forked twig placed so as to press lightly on the underside of a young / footstalk caused it, in 12 hrs, to bend greatly, and ultimately to such an extent that the leaf passed to the opposite side of the stem; the forked stick having been removed, the leaf slowly recovered its former position.

The young leaves spontaneously and gradually change their position: when first developed the petioles are upturned and parallel to the stem; they then slowly bend downwards, remaining for a short time at right angles to the stem, and then become so much arched downwards that the blade of the leaf points to the ground with its tip curled inwards, so that the whole petiole and leaf together form a hook. They are thus enabled to catch hold of any twig with which they may be brought into contact by the revolving movement of the internodes. If this does not happen, they retain their hooked shape for a considerable time, and then bending upwards reassume their original upturned / position, which is preserved ever afterwards. The petioles which have clasped any object soon become much thickened and strengthened, as may be seen in the drawing.

Clematis montana. The long, thin petioles of the leaves, whilst young, are sensitive, and when lightly rubbed bend to the rubbed side, subsequently becoming straight. They are far more sensitive than the petioles of *C. glandulosa*; for a loop of thread weighing a quarter of a grain (16·2 mg) caused them to bend; a loop weighing only one-eighth of a grain (8·1 mg) sometimes acted and sometimes did not act. The sensitiveness extends from the blade of the leaf to the stem. I may here state that I ascertained in all cases the weights of the string and thread used by carefully weighing 50 inches in a chemical balance, and then cutting off measured lengths. The main petiole carries three leaflets; but their short, sub-petioles are not sensitive. A young, inclined shoot (the plant being in the greenhouse) made a large circle opposed to the course of the sun in 4 hrs 20 m, but the next day, being very cold, the time was 5 hrs 10 m. A stick placed near a revolving stem was soon struck by the petioles which stand out at right angles, and the revolving movement was thus arrested. The petioles then began, being excited by the contact, to slowly wind round the stick. When the stick was thin, a petiole sometimes wound twice round it. The opposite leaf was in no way affected. The attitude assumed by the stem after the petiole had / clasped the stick, was that of a man standing by a column, who throws his arm horizontally round it. With respect to the stem's power of twining, some remarks will be made under *C. calycina*.

Clematis Sieboldi. A shoot made three revolutions against the sun at an average rate of 3 hrs 11 m. The power of twining is like that of the last species. Its leaves are nearly similar in structure and in function, excepting that the sub-petioles of the lateral and terminal leaflets are sensitive. A loop of thread, weighing one-eighth of a grain, acted on the main petiole, but not until two or three days had elapsed. The leaves have the remarkable habit of spontaneously revolving, generally in vertical ellipses, in the same manner, but in a less degree, as will be described under *C. microphylla*.

Clematis calycina. The young shoots are thin and flexible: one revolved, describing a broad oval, in 5 hrs 30 m, and another in 6 hrs 12 m. They followed the course of the sun; but the course, if observed long enough, would probably be found to vary in this species, as well as in all the others of the genus. It is a rather better twiner than the two last species: the stem sometimes made two spiral turns round a thin stick, if free from twigs; it then ran straight up for a space, and reversing its course took one or two turns in an opposite direction. This reversal of the spire occurred in all the foregoing species. The leaves are so small compared with those of most of the other species, that the petioles at first seem ill-adapted for clasping. / Nevertheless, the main service of the revolving movement is to bring them into contact with surrounding objects, which are slowly but securely seized. The young petioles, which alone are sensitive, have their ends bowed a little downwards, so as to be in a slight degree hooked; ultimately the whole leaf, if it catches nothing, becomes level. I gently rubbed with a thin twig the lower surfaces of two young petioles; and in 2 hrs 30 m they were slightly curved downwards; in 5 hrs, after being rubbed, the end of one was bent completely back, parallel to the basal portion; in 4 hrs subsequently it became nearly straight again. To show how sensitive the young petioles are, I may mention that I just touched the undersides of two with a little water-colour, which when dry formed an excessively thin and minute crust; but this sufficed in 24 hrs to cause both to bend downwards. Whilst the plant is young, each leaf consists of three divided leaflets, which barely have distinct petioles, and these are not sensitive; but when the plant is well grown, the petioles of the two lateral and terminal leaflets are of considerable length, and become sensitive so as to be capable of clasping an object in any direction.

When a petiole has clasped a twig, it undergoes some remarkable changes, which may be observed with the other species, but in a less strongly marked manner, and will here be described once for all. The clasped petiole in the course of two or three days swells greatly, and ultimately becomes nearly twice as / thick as the opposite one which has clasped nothing. When thin transverse slices of the two are placed under the microscope their difference is conspicuous: the side of the petiole which has been in contact with the support, is formed of a layer of colourless cells with their longer axes directed from the centre, and these are very much larger than the corresponding cells in the opposite or unchanged petiole; the central cells, also, are in some degree enlarged, and the whole is much indurated. The exterior surface generally becomes bright red. But a far greater change takes place in the nature of the tissues than that which is visible: the petiole of the unclasped leaf is flexible and can be snapped easily, whereas the clasped one acquires an extraordinary degree of toughness and rigidity, so that considerable force is required to pull it into pieces. With this change, great durability is probably acquired; at least this is the case with the clasped petioles of *Clematis vitalba*. The meaning of these changes is obvious, namely, that the petioles may firmly and durably support the stem.

Clematis microphylla, var. *leptophylla*. The long and thin internodes of this Australian species revolve sometimes in one direction and sometimes in an opposite one, describing long, narrow, irregular ellipses or large circles. Four revolutions were completed within five minutes of the same average rate of 1 hr 51 m; so that this species moves more quickly than the others of the genus. The shoots, when placed near a vertical stick, either twine round it, or clasp it / with the basal portions of their petioles. The leaves whilst young are nearly of the same shape as those of *C. viticella*, and act in the same manner like a hook, as will be described under that species. But the leaflets are more divided, and each segment whilst young terminates in a hardish point, which is much curved downwards and inwards; so that the whole leaf readily catches hold of any neighbouring object. The petioles of the young terminal leaflets are acted on by loops of thread weighing ⅛th and even ¹⁄₁₆th of a grain. The basal portion of the main petiole is much less sensitive, but will clasp a stick against which it presses.

The leaves, whilst young, are continually and spontaneously moving slowly. A bell-glass was placed over a shoot secured to a stick, and the movements of the leaves were traced on it during several days. A very irregular line was generally formed; but one day, in the course of eight hours and three-quarters, the figure clearly represented three and a half irregular ellipses, the most perfect one of which was completed in 2 hrs 35 m. The two opposite leaves moved independently of each other. This movement of the leaves would aid that of the internodes in bringing the petioles into contact with surrounding objects. I discovered this movement too late to be enabled to observe it in the other species; but from analogy I can hardly doubt that the leaves of at least *C. viticella*, *C. flammula*, and *C. vitalba* move spontaneously; and, judging from *C. Sieboldi*, this probably is the case with / *C. montana* and

C. calycina. I ascertained that the simple leaves of *C. glandulosa* exhibited no spontaneous revolving movement.

Clematis viticella, var. *venosa*. In this and the two following species the power of spirally twining is completely lost, and this seems due to the lessened flexibility of the internodes and to the interference caused by the large size of the leaves. But the revolving movement, though restricted, is not lost. In our present species a young internode, placed in front of a window, made three narrow ellipses, transversely to the direction of the light, at an average rate of 2 hrs 40 m. When placed so that the movements were to and from the light, the rate was greatly accelerated in one half of the course, and retarded in the other, as with twining plants. The ellipses were small; the longer diameter, described by the apex of a shoot bearing a pair of not expanded leaves, was only $4\frac{5}{8}$ inches, and that by the apex of the penultimate internode only $1\frac{1}{8}$ inch. At the most favourable period of growth each leaf would hardly be carried to and fro by the movement of the internodes more than two or three inches, but, as above stated, it is probable that the leaves themselves move spontaneously. The movement of the whole shoot by the wind and by its rapid growth, would probably be almost equally efficient as these spontaneous movements, in bringing the petioles into contact with surrounding objects.

The leaves are of large size. Each bears three pairs of lateral leaflets and a terminal one, all supported on / rather long sub-petioles. The main petiole bends a little angularly downwards at each point where a pair of leaflets arises (see fig. 2), and the petiole of the terminal leaflet is bent downwards at right angles; hence the whole petiole, with its rectangularly bent extremity, acts as a hook. This hook, the lateral petioles being directed a little upwards, forms an excellent grappling apparatus, by which the leaves readily become entangled with surrounding objects. If they catch nothing, the whole petiole ultimately grows straight. The main petiole, the sub-petioles, and the three branches into which each basi-lateral sub-petiole is generally subdivided, are all sensitive. The basal portion of the main petiole, between the stem and the first pair of leaflets, is less sensitive than the remainder; it will, however, clasp a stick / with which it is left in contact. The inferior surface of the rectangularly bent terminal portion (carrying the terminal leaflet), which forms the inner side of the end of the hook, is the most sensitive part; and this portion is manifestly best adapted to catch a distant support. To show the difference in sensibility, I gently placed loops of string of the same weight (in one instance weighing only 0·82 of a grain or 53·14 mg) on the several lateral sub-petioles and on the terminal one; in a few hours the latter was bent, but after 24 hrs no effect was produced on the other sub-petioles. Again, a terminal sub-petiole placed in contact with a thin stick became sensibly curved in 45 m, and in 1 hr 10 m moved through ninety degrees; whilst a lateral sub-petiole did not become sensibly curved until 3 hrs 30 m had elapsed. In all cases, if the sticks are taken away, the petioles continue to move during many hours afterwards; so they do after a slight rubbing; but they become straight again, after about a day's interval, that is if the flexure has not been very great or long continued.

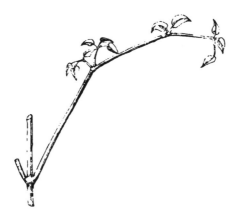

Fig. 2   A young leaf of *Clematis viticella*

The graduated difference in the extension of the sensitiveness in the petioles of the above-described species deserves notice. In *C. montana* it is confined to the main petiole, and has not spread to the sub-petioles of the three leaflets; so it is with young plants of *C. calycina*, but in older plants it spreads to the three sub-petioles. In *C. viticella* the sensitiveness has spread to the petioles of the seven leaflets, and to the subdivisions of the basi-lateral sub-petioles. But in / this latter species it has diminished in the basal part of the main petiole, in which alone it resided in *C. montana*; whilst it has increased in the abruptly bent terminal portion.

Clematis flammula. The rather thick, straight, and stiff shoots, whilst

37

growing vigorously in the spring, make small oval revolutions, following the sun in their course. Four were made at an average rate of 3 hrs 45 m. The longer axis of the oval, described by the extreme tip, was directed at right angles to the line joining the opposite leaves; its length was in one case only 1⅜, and in another case 1⅝ inch; so that the young leaves were moved a very short distance. The shoots of the same plant observed in midsummer, when growing not so quickly, did not revolve at all. I cut down another plant in the early summer, so that by 1 August it had formed new and moderately vigorous shoots; these, when observed under a bell-glass, were on some days quite stationary, and on other days moved to and fro only about the eighth of an inch. Consequently the revolving power is much enfeebled in this species, and under unfavourable circumstances is completely lost. The shoot must depend for coming into contact with surrounding objects on the probable, though not ascertained spontaneous movement of the leaves, on rapid growth, and on movement from the wind. Hence, perhaps, it is that the petioles have acquired a high degree of sensitiveness as a compensation for the little power of movement in the shoots.

The petioles are bowed downwards, and have the / same general hook-like form as in *C. viticella*. The medial petiole and the lateral sub-petioles are sensitive, especially the much bent terminal portion. As the sensitiveness is here greater than in any other species of the genus observed by me, and is in itself remarkable, I will give fuller details. The petioles, when so young that they have not separated from one another, are not sensitive; when the lamina of a leaflet has grown to a quarter of an inch in length (that is, about one-sixth of its full size), the sensitiveness is highest; but at this period the petioles are relatively much more fully developed than are the blades of the leaves. Full-grown petioles are not in the least sensitive. A thin stick placed so as to press lightly against a petiole, having a leaflet a quarter of an inch in length, caused the petiole to bend in 3 hrs 15 m. In another case a petiole curled completely round a stick in 12 hrs. These petioles were left curled for 24 hrs, and the sticks were then removed; but they never straightened themselves. I took a twig, thinner than the petiole itself, and with it lightly rubbed several petioles four times up and down; these in 1 hr 45 m became slightly curled; the curvature increased during some hours and then began to decrease, but after 25 hrs from the time of rubbing a vestige of the curvature remained. Some other petioles similarly rubbed twice, that is, once up and once down, became perceptibly curved in about 2 hrs 30 m, the terminal sub-petiole moving more than the lateral sub-petioles; they all became straight again in between 12 hrs and 14 hrs. Lastly, a / length of about one-eighth of an inch of a sub-petiole, was lightly rubbed with the same twig only once; it became slightly curved in 3 hrs, remaining so during 11 hrs, but by the next morning was quite straight.

The following observations are more precise. After trying heavier pieces of string and thread, I placed a loop of fine string, weighing 1·04 gr (67·4 mg) on a terminal sub-petiole: in 6 hrs 40 m a curvature could be seen; in 24 hrs the petiole formed an open ring round the string; in 48 hrs the ring had almost closed on the string, and in 72 hrs seized it so firmly, that some force was necessary for its withdrawal. A loop weighing 0·52 of a grain (33·7 mg) caused in 14 hrs a lateral sub-petiole just perceptibly to curve, and in 24 hrs it moved through ninety degrees. These observations were made during the summer: the following were made in the spring, when the petioles apparently are more sensitive: a loop of thread, weighing one-eighth of a grain (8·01 mg), produced no effect on the lateral sub-petioles, but placed on a terminal one, caused it, after 24 hrs, to curve moderately; the curvature, though the loop remained suspended, was after 48 hrs diminished, but never disappeared; showing that the petiole had become partially accustomed to the insufficient stimulus. This experiment was twice repeated with nearly the same result. Lastly, a loop of thread, weighing only one-sixteenth of a grain (4·05 mg) was twice gently placed by a forceps on a terminal sub-petiole (the plant being, of course, in a still and closed room), and this weight certainly caused a flexure, which very / slowly increased until the petiole moved through nearly ninety degrees: beyond this it did not move; nor did the petiole, the loop remaining suspended, ever become perfectly straight again.

When we consider, on the one hand, the thickness and stiffness of the petioles, and, on the other hand, the thinness and softness of fine cotton thread, and what an extremely small weight one-sixteenth of a grain (4·05 mg) is, these facts are remarkable. But I have reason to believe that even a less weight excites curvature when pressing over a broader surface than that acted on by a thread. Having noticed that the end of a suspended string which accidentally touched a petiole, caused it to bend, I took two pieces of thin twine, 10 inches in length (weighing 1·64 gr), and, tying them to a stick, let them hang as nearly perpendicularly downwards as their thinness and flexuous form, after being stretched, would permit; I then quietly placed their ends so as just to rest on two petioles, and these certainly became curved in 36 hrs. One of the ends touched the angle between a terminal and lateral sub-petiole, and it was in 48 hours caught between them as by a forceps. In these cases the pressure, though spread over a wider surface than that touched by the cotton thread, must have been excessively slight.

Clematis vitalba. The plants were in pots and not healthy, so that I dare not trust my observations, which indicate much similarity in habits with *C. flammula.* I mention this species only because I have seen many / proofs that the petioles in a state of nature are excited to movement by very slight pressure. For instance, I have found them embracing thin withered blades of grass, the soft young leaves of a maple, and the flower-peduncles of the quaking-grass or Briza. The latter are about as thick as the hair of a man's beard, but they were completely surrounded and clasped. The petioles of a leaf, so young that none of the leaflets were expanded, had partially seized a twig. Those of almost all the old leaves, even when unattached to any object, are much convoluted; but this is owing to their having come, whilst young, into contact during several hours with some object subsequently removed. With none of the above-described species, cultivated in pots and carefully observed, was there any permanent bending of the petioles without the stimulus of contact. In winter, the blades of the leaves of *C. vitalba* drop off; but the petioles (as was observed by Mohl) remain attached to the branches, sometimes during two seasons; and, being convoluted, they curiously resemble true tendrils, such as those possessed by the allied genus *Naravelia.* The petioles which have clasped some object become much more stiff, hard, and polished than those which have failed in this their proper function.

TROPAEOLUM

I observed *T. tricolorum, T. azureum, T. pentaphyllum, T. peregrinum, T. elegans, T. tuberosum,* and a dwarf variety of, as I believe, *T. minus.*

Tropaeolum tricolorum, var. *grandiflorum.* The flexible shoots, which first rise from the tubers, are / as thin as fine twine. One such shoot revolved in a course opposed to the sun, at an average rate, judging from three revolutions, of 1 hr 23 m; but no doubt the direction of the revolving movement is variable. When the plants have grown tall and are branched, all the many lateral shoots revolve. The stem, whilst young, twines regularly round a thin vertical stick, and in one case I counted eight spiral turns in the same direction; but when grown older, the stem often runs straight up for a space, and, being arrested by the clasping petioles, makes one or two spires in a reversed direction. Until the plant grows to a height of two or three feet, requiring about a month from the time when the first shoot appears above ground, no true leaves are produced, but, in their place, filaments coloured like the stem. The extremities of these filaments are pointed, a little flattened, and furrowed on the upper surface. They never become developed into leaves. As the plant grows in height new filaments are produced with slightly enlarged tips; then others, bearing on each side of the enlarged medial tip a rudimentary segment of a leaf; soon other segments appear, and at last a perfect leaf is formed, with seven deep segments. So that on the same plant we may see every step, from tendril-like clasping filaments to perfect leaves with clasping petioles. After the plant has grown to a considerable height, and is secured to its support by the petioles of the true leaves, the clasping filaments on the lower part of the stem wither and drop off; so that they perform only a temporary service. /

These filaments or rudimentary leaves, as well as the petioles of the perfect leaves, whilst young, are highly sensitive on all sides to a touch. The slightest rub caused them to curve towards the rubbed side in about three minutes, and one bent itself into a ring in six minutes; they subsequently became straight. When, however, they have once completely clasped a stick, if this is removed, they do not straighten themselves. The most remarkable fact, and one which I have observed in no other species of the genus, is that the filaments and the petioles of the young leaves, if they catch no object, after standing for some days in their original position, spontaneously and slowly oscillate a little from side to side, and then move towards the stem and clasp it. They likewise often become, after a time, in some degree spirally contracted. They therefore fully deserve to be called tendrils, as they are used for climbing, are sensitive to a touch, move spontaneously, and ultimately contract into a spire, though an imperfect one. The present species would have been classed among the tendril-bearers, had not these characters been confined to early youth. During maturity it is a true leaf-climber.

Tropaeolum azureum. An upper internode made four revolutions,

41

following the sun, at an average rate of 1 hr 47 m. The stem twined spirally round a support in the same irregular manner as that of the last species. Rudimentary leaves or filaments do not exist. The petioles of the young leaves are very sensitive: a single light rub with a twig caused one / to move perceptibly in 5 m, and another in 6 m. The former became bent at right angles in 15 min, and became straight again in between 5 hrs and 6 hrs. A loop of thread weighing ⅛th of a grain caused another petiole to curve.

Tropaeolum pentaphyllum. This species has not the power of spirally twining, which seems due, not so much to a want of flexibility in the stem, as to continual interference from the clasping petioles. An upper internode made three revolutions, following the sun, at an average rate of 1 hr 46 m. The main purpose of the revolving movement in all the species of *Tropaeolum* manifestly is to bring the petioles into contact with some supporting object. The petiole of a young leaf, after a slight rub, became curved in 6 m; another, on a cold day, in 20 m, and others in from 8 m to 10 m. Their curvature usually increased greatly in from 15 m to 20 m, and they became straight again in between 5 hrs and 6 hrs, but on one occasion in 3 hrs. When a petiole has fairly clasped a stick, it is not able, on the removal of the stick, to straighten itself. The free upper part of one, the base of which had already clasped a stick, still retained the power of movement. A loop of thread weighing ⅛th of a grain caused a petiole to curve; but the stimulus was not sufficient, the loop remaining suspended, to cause a permanent flexure. If a much heavier loop be placed in the angle between the petiole and the stem, it produces no effect; whereas we have seen with *Clematis montana* that the angle between the stem and petiole is sensitive. /

Tropaeolum peregrinum. The first-formed internodes of a young plant did not revolve, resembling in this respect those of a twining plant. In an older plant the four upper internodes made three irregular revolutions, in a course opposed to the sun, at an average rate of 1 hr 48 min. It is remarkable that the average rate of revolution (taken, however, but from few observations) is very nearly the same in this and the two last species, namely, 1 hr 47 m, 1 hr 46 m, and 1 hr 48 m. The present species cannot twine spirally, which seems mainly due to the rigidity of the stem. In a very young plant, which did not revolve, the petioles were not sensitive. In older plants the petioles of quite young leaves, and of leaves as much as an inch and a quarter in diameter, are sensitive. A moderate rub caused one to curve in 10 m, and others in 20 m. They became straight again in between 5 hrs 45 m and 8 hrs. Petioles which have naturally come into contact with a stick, sometimes take two turns round it. After they have clasped a support, they become rigid and hard. They are less sensitive to a weight than in the previous species; for loops of string weighing 0·82 of a grain (53·14 mg), did not cause any curvature, but a loop of double this weight (1·65 gr) acted.

Tropaeolum elegans. I did not make many observations on this species. The short and stiff internodes revolve irregularly, describing small oval figures. One oval was completed in 3 hrs. A young petiole, when rubbed, became slightly curved in 17 m; and / afterwards much more so. It was nearly straight again in 8 hrs.

Tropaeolum tuberosum. On a plant nine inches in height, the internodes did not move at all; but on an older plant they moved irregularly and made small imperfect ovals. These movements could be detected only by being traced on a bell-glass placed over the plant. Sometimes the shoots stood still for hours; during some days they moved only in one direction in a crooked line; on other days they made small irregular spires or circles, one being completed in about 4 hrs. The extreme points reached by the apex of the shoot were only about one or one and a half inches asunder; yet this slight movement brought the petioles into contact with some closely surrounding twigs, which were then clasped. With the lessened power of spontaneously revolving, compared with that of the previous species, the sensitiveness of the petioles is also diminished. These, when rubbed a few times, did not become curved until half an hour had elapsed; the curvature increased during the next two hours, and then very slowly decreased; so that they sometimes required 24 hrs to become straight again. Extremely young leaves have active petioles; one with the lamina only 0·15 of an inch in diameter, that is, about a twentieth of the full size, firmly clasped a thin twig. But leaves grown to a quarter of their full size can likewise act.

Tropaeolum minus (?). The internodes of a variety named 'dwarf crimson Nasturtium' did not revolve, / but moved in a rather irregular course during the day to the light, and from the light at night. The petioles, when well rubbed, showed no power of curving; nor could I see that they ever clasped any neighbouring object. We have seen in this genus a gradation from species such as *T. tricolorum*, which have extremely sensitive petioles, and internodes which rapidly revolve and spirally twine up a support, to other species such as *T. elegans* and *T. tuberosum*, the petioles of which are much less sensitive, and the internodes of which have very feeble revolving powers and cannot spirally twine round a support, to this last species, which has entirely lost or never acquired these faculties. From the general character of the genus, the loss of power seems the more probable alternative.

In the present species, in *T. elegans*, and probably in others, the flower-peduncle, as soon as the seed-capsule begins to swell, spontaneously bends abruptly downwards and becomes somewhat convoluted. If a stick stands in the way, it is to a certain extent clasped; but, as far as I have been able to observe, this clasping movement is independent of the stimulus from contact.

ANTIRRHINEAE

In this tribe (Lindley) of the Scrophulariaceae, at least four of the seven included genera have leaf-climbing species.

Maurandia Barclayana. A thin, slightly bowed shoot made two revolutions, following the sun, each in 3 hrs 17 min; on the previous day this same shoot revolved in an opposite direction. The shoots do not twine spirally, but climb excellently by the aid of / their young and sensitive petioles. These petioles, when lightly rubbed, move after a considerable interval of time, and subsequently become straight again. A loop of thread weighing ⅛th of a grain caused them to bend.

Maurandia semperflorens. This freely growing species climbs exactly like the last, by the aid of its sensitive petioles. A young internode made two circles, each in 1 hr 46 min; so that it moved almost twice as rapidly as the last species. The internodes are not in the least sensitive to a touch or pressure. I mention this because they are sensitive in a closely allied genus, namely, Lophospermum. The present species is unique in one respect. Mohl asserts (p. 45) that 'the flower-peduncles, as well as the petioles, wind like tendrils'; but he classes as tendrils such objects as the spiral flower-stalks of the *Vallisneria*. This remark, and the fact of the flower-peduncles being decidedly flexuous, led me carefully to examine them. They never act as true tendrils; I repeatedly placed thin sticks in contact with young and old peduncles, and I allowed nine vigorous plants to grow through an entangled mass of branches; but in no one instance did they bend round any object. It is indeed in the highest degree improbable that this should occur, for they are generally developed on branches which have already securely clasped a support by the petioles of their leaves; and when borne on a free depending branch, they are not produced by the terminal portion of the internode / which alone has the power of revolving; so that they could be brought only by accident into contact with any neighbouring object. Nevertheless (and this is the remarkable fact) the flower-peduncles, whilst young, exhibit feeble revolving powers, and are slightly sensitive to a touch. Having selected some stems which had firmly clasped a stick by their petioles, and having placed a bell-glass over them, I traced the movements of the young flower-peduncles. The tracing generally formed a short and extremely irregular line, with little loops in its course. A young peduncle 1½ inch in length was carefully observed during a whole day, and it made four and a half narrow, vertical, irregular, and short ellipses – each at an average rate of about 2 hrs 25 m. An adjoining peduncle described during the same time similar, though fewer, ellipses. As the plant had occupied for some time exactly the same position, these movements could not be attributed to any change in the action of the light. Peduncles, old enough for the coloured petals to be just visible, do not move. With respect to irritability,[1] I rubbed two young peduncles (1½ inch in length) a few times very lightly with a thin twig; one was rubbed on the upper, and the other on the lower side, and they became in between 4 hrs and 5 hrs distinctly bowed towards / these sides; in 24 hrs subsequently, they straightened themselves. Next day they were rubbed on the opposite sides, and they became perceptibly curved towards these sides. Two other and younger peduncles (three-fourths of an inch in length) were lightly rubbed on their adjoining sides, and they became so much curved towards one another, that the arcs of the bows stood at nearly right angles to their previous direction; and this was the greatest movement seen by me. Subsequently they straight-ened themselves. Other peduncles, so young as to be only three-tenths of an inch in length, became curved when rubbed. On the other hand, peduncles above 1½ inch in length required to be rubbed two or three times, and then became only just perceptibly bowed. Loops of thread

---

[1] It appears from A. Kerner's interesting observations, that the flower-peduncles of a large number of plants are irritable, and bend when they are rubbed or shaken: *Die Schutzmittel des Pollens*, 1873, p. 34.

suspended on the peduncles produced no effect; loops of string, however, weighing 0·82 and 1·64 of a grain sometimes caused a slight curvature; but they were never closely clasped, as were the far lighter loops of thread by the petioles.

In the nine vigorous plants observed by me, it is certain that neither the slight spontaneous movements nor the slight sensitiveness of the flower-peduncles aided the plants in climbing. If any member of the Scrophulariaceae had possessed tendrils produced by the modification of flower-peduncles, I should have thought that this species of *Maurandia* had perhaps retained a useless or rudimentary vestige of a former habit; but this view cannot be maintained. We may suspect that, owing to the principle of correlation, / the power of movement has been transferred to the flower-peduncles from the young internodes, and sensitiveness from the young petioles. But to whatever cause these capacities are due, the case is interesting; for, by a little increase in power through Natural Selection, they might easily have been rendered as useful to the plant in climbing, as are the flower-peduncles (hereafter to be described) of Vitis or Cardiospermum.

Rhodochiton volubile. A long flexible shoot swept a large circle, following the sun, in 5 hrs 30 m; and, as the day became warmer, a second circle was completed in 4 hrs 10 m. The shoots sometimes make a whole or a half spire round a vertical stick, they then run straight up for a space, and afterwards turn spirally in an opposite direction. The petioles of very young leaves about one-tenth of their full size, are highly sensitive, and bend towards the side which is touched; but they do not move quickly. One was perceptibly curved in 1 hr 10 m, after being lightly rubbed, and became considerably curved in 5 hrs 40 m; some others were scarcely curved in 5 hrs 30 m, but distinctly so in 6 hrs 30 m. A curvature was perceptible in one petiole in between 4 hrs 30 m and 5 hrs, after the suspension of a little loop of string. A loop of fine cotton thread, weighing one sixteenth of a grain (4·06 mg), not only caused a petiole slowly to bend, but was ultimately so firmly clasped that it could be withdrawn only by some little force. The petioles, when coming into contact with a stick, take / either a complete or half a turn round it, and ultimately increase much in thickness. They do not possess the power of spontaneously revolving.

Lophospermum scandens, var. *purpureum.* Some long, moderately thin internodes made four revolutions at an average rate of 3 hrs 15 m. The course pursued was very irregular, namely, an extremely narrow ellipse, a large circle, an irregular spire or a zigzag line, and sometimes the apex stood still. The young petioles, when brought by the revolving movement into contact with sticks, clasped them, and soon increased considerably in thickness. But they are not quite so sensitive to a weight as those of the *Rhodochiton*, for loops of thread weighing one-eighth of a grain did not always cause them to bend.

This plant presents a case not observed by me in any other leaf-climber or twiner,[2] namely, that the young internodes of the stem are sensitive to a touch. When a petiole of this species clasps a stick, it draws the base of the internode against it; and then the internode itself bends towards the stick, which is caught between the stem and the petiole as by a pair of pincers. The internode afterwards straightens itself, excepting the part in actual contact with the stick. Young internodes alone are sensitive, and these are sensitive on all sides along their whole length. I made / fifteen trials by twice or thrice lightly rubbing with a thin twig several internodes; and in about 2 hrs, but in one case in 3 hrs, all were bent: they became straight again in about 4 hrs afterwards. An internode, which was rubbed as often as six or seven times, became just perceptibly curved in 1 hr 15 m, and in 3 hrs the curvature increased much; it became straight again in the course of the succeeding night. I rubbed some internodes one day on one side, and the next day either on the opposite side or at right angles to the first side; and the curvature was always towards the rubbed side.

According to Palm (p. 63), the petioles of *Linaria cirrhosa* and, to a limited degree, those of *L. elatine* have the power of clasping a support.

SOLANACEAE

Solanum jasminoides. Some of the species in this large genus are twiners; but the present species is a true leaf-climber. A long, nearly upright shoot made four revolutions, moving against the sun, very regularly at an average rate of 3 hrs 26 m. The shoots, however, sometimes stood still. It is considered a greenhouse plant; but when kept there, the petioles took several days to clasp a stick: in the hothouse a stick was clasped in 7 hrs. In the greenhouse, a petiole was not affected by a loop of string, suspended during several days and weighing 2½ grains (163

---

[2] I have already referred to the case of the twining stem of Cuscuta, which, according to H. de Vries (ibid., p. 322) is sensitive to a touch like a tendril.

47

mg); but in the hothouse one was made to curve by a loop weighing 1·64 gr (106·27 mg); and, on the removal of the string, it became straight again. Another petiole was not at all acted on by a loop / weighing only 0·82 of a grain (53·14 mg). We have seen that the petioles of some other leaf-climbing plants are affected by one-thirteenth of this latter weight. In this species, and in no other leaf-climber seen by me, a full-grown leaf is capable of clasping a stick; but

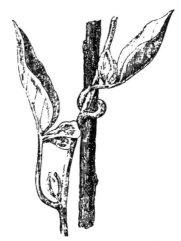

Fig. 3 *Solanum jasminoides*, with one of its petioles clasping a stick

in the greenhouse the movement was so extraordinarily slow that the act required several weeks; on each succeeding week it was clear that the petiole had become more and more curved, until at last it firmly clasped the stick.

The flexible petiole of a half or a quarter grown leaf which has clasped an object for three or four days increases much in thickness, and after several weeks becomes so wonderfully hard and rigid that it / can hardly be removed from its support. On comparing a thin transverse slice of such a petiole with one from an older leaf growing close beneath, which had not clasped anything, its diameter was found to be fully doubled, and its structure greatly changed. In two other petioles similarly compared, and here represented, the increase in diameter was not quite so great. In the section of the petiole in its ordinary state (A), we see a semilunar band of cellular tissue (not well shown in the woodcut) differing slightly in appearance from that outside it, and including three closely approximate groups of dark

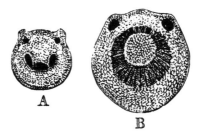

Fig. 4   *Solanum jasminoides*
A. Section of a petiole in its ordinary state
B. Section of a petiole some weeks after it had clasped a stick, as shown in Fig. 3

vessels. Near the upper surface of the petiole, beneath two exterior ridges, there are two other small circular groups of vessels. In the section of the petiole (B) which had clasped during several weeks a stick, the two exterior ridges have become much less prominent, and the two groups of woody vessels beneath them much increased in diameter. The semilunar band has been converted into a complete ring of very hard, white, woody / tissue, with lines radiating from the centre. The three groups of vessels, which, though near together, were before distinct, are now completely blended. The upper part of this ring of woody vessels, formed by the prolongation of the horns of the original semilunar band, is narrower than the lower part, and slightly less compact. This petiole after clasping the stick had actually become thicker than the stem from which it arose; and this was chiefly due to the increased thickness of the ring of wood. This ring presented, both in a transverse and longitudinal section, a closely similar structure to that of the stem. It is a singular morphological fact that the petiole should thus acquire a structure almost identically the same with that of the axis; and it is a still more singular physiological fact that so great a change should have been induced by the mere act of clasping a support.[3]

[3] Dr Maxwell Masters informs me that in almost all petioles which are cylindrical, such as those bearing peltate leaves, the woody vessels form a closed ring; semilunar bands of vessels being confined to petioles which are channelled along their upper surfaces. In accordance with this statement, it may be observed that the enlarged and clasped petiole of the *Solanum*, with its closed ring of woody vessels, has become more cylindrical than it was in its original unclasped condition.

FUMARIACEAE

Fumaria officinalis. It could not have been anticipated that so lowly a plant as this Fumaria should have been a climber. It climbs by the aid of the main and lateral petioles of its compound leaves; and even the much-flattened terminal / portion of the petiole can seize a support. I have seen a substance as soft as a withered blade of grass caught. Petioles which have clasped any object ultimately become rather thicker and more cylindrical. On lightly rubbing several petioles with a twig, they became perceptibly curved in 1 hr 15 m, and subsequently straightened themselves. A stick gently placed in the angle between two sub-petioles excited them to move, and was almost clasped in 98 hrs. A loop of thread, weighing one-eighth of a grain, caused, after 12 hrs and before 20 hrs had elapsed, a considerable curvature; but it was never fairly clasped by the petiole. The young internodes are in continual movement, which is considerable in extent, but very irregular; a zigzag line, or a spire crossing itself, or a figure of 8 being formed. The course during 12 hrs, when traced on a bell-glass, apparently represented about four ellipses. The leaves themselves likewise move spontaneously, the main petioles curving themselves in accordance with the movements of the internodes; so that when the latter moved to one side, the petioles moved to the same side, then, becoming straight, reversed their curvature. The petioles, however, do not move over a wide space, as could be seen when a shoot was securely tied to a stick. The leaf in this case followed an irregular course, like that made by the internodes.

Adlumia cirrhosa. I raised some plants late in the summer; they formed very fine leaves, but threw up no central stem. The first-formed leaves were not / sensitive; some of the later ones were so, but only towards their extremities, which were thus enabled to clasp sticks. This could be of no service to the plant, as these leaves rose from the ground; but it showed what the future character of the plant would have been, had it grown tall enough to climb. The tip of one of these basal leaves, whilst young, described in 1 hr 36 m a narrow ellipse, open at one end, and exactly three inches in length; a second ellipse was broader, more irregular, and shorter, viz., only 2½ inches in length, and was completed in 2 hrs 2 m. From the analogy of *Fumaria* and *Corydalis*, I have no doubt that the internodes of Adlumia have the power of revolving.

Corydalis claviculata. This plant is interesting from being in a condition so exactly intermediate between a leaf-climber and a tendril-bearer, that it might have been described under either head; but, for reasons hereafter assigned, it has been classed among tendril-bearers.

Besides the plants already described, *Bignonia unguis* and its close allies, though aided by tendrils, have clasping petioles. According to Mohl (p. 40), *Cocculus Japonicus* (one of the Menispermaceae) and a fern, the *Ophioglossum Japonicum* (p. 39), climb by their leafstalks.

We now come to a small section of plants which climb by means of the produced midribs or tips of their leaves. /

LILIACEAE

Gloriosa Plantii. The stem of a half-grown plant continually moved, generally describing an irregular spire, but sometimes oval figures with the longer axes directed in different lines. It either followed the sun, or moved in an opposite course, and sometimes stood still before reversing its direction. One oval was completed in 3 hrs 40 m; of two horseshoe-shaped figures, one was completed in 4 hrs 35 m and the other in 3 hrs. The shoots, in their movements, reached points between four and five inches asunder. The young leaves, when first developed, stand up nearly vertically; but by the growth of the axis, and by the spontaneous bending down of the terminal half of the leaf, they soon become much inclined, and ultimately horizontal. The end of the leaf forms a narrow, ribbon-like, thickened projection, which at first is nearly straight, but by the time the leaf gets into an inclined position, the end bends downwards into a well-formed hook. This hook is now strong and rigid enough to catch any object, and, when caught, to anchor the plant and stop the revolving movement. Its inner surface is sensitive, but not in nearly so high a degree as that of the many before-described petioles; for a loop of string, weighing 1·64 grain, produced no effect. When the hook has caught a thin twig or even a rigid fibre, the point may be perceived in from 1 hr to 3 hrs to have curled a little inwards; and, under favourable circumstances, it curls round and permanently seizes an object in from 8 hrs to 10 hrs. / The hook when first formed, before the leaf has bent downwards, is but little sensitive. If it catches hold of nothing, it remains open and sensitive for a long time; ultimately the extremity spontaneously and slowly curls inwards, and makes a button-like, flat, spiral coil at the end of the leaf. One leaf was watched, and the hook remained open for thirty-three days; but during the last week the tip had curled so much inwards that only a very thin twig could have been inserted within it. As soon as the tip has curled so much inwards that the hook is converted into a ring, its sensibility is lost; but as long as it remains open some sensibility is retained.

Whilst the plant was only about six inches in height, the leaves, four or five in number, were broader than those subsequently produced; their soft and but little-attenuated tips were not sensitive, and did not form hooks; nor did the stem then revolve. At this early period of growth, the plant can support itself; its climbing powers are not required, and consequently are not developed. So again, the leaves on the summit of a full-grown flowering plant, which would not require to climb any higher, were not sensitive and could not clasp a stick. We thus see how perfect is the economy of nature.

COMMELYNACEAE

Flagellaria Indica. From dried specimens it is manifest that this plant climbs exactly like the Gloriosa. A young plant 12 inches in height, and bearing fifteen leaves, had not a single leaf as yet produced into a hook or tendril-like filament; nor did / the stem revolve. Hence this plant acquires its climbing powers later in life than does the Gloriosa lily. According to Mohl (p. 41), Uvularia (Melanthaceae) also climbs like Gloriosa.

These three last-named genera are monocotyledons; but there is one dicotyledon, namely Nepenthes, which is ranked by Mohl (p. 41) among tendril-bearers; and I hear from Dr Hooker that most of the species climb well at Kew. This is effected by the stalk or midrib between the leaf and the pitcher coiling round any support. The twisted part becomes thicker; but I observed in Mr Veitch's hothouse that the stalk often takes a turn when not in contact with any object, and that this twisted part is likewise thickened. Two vigorous young plants of *N. laevis* and *N. distillatoria*, in my hothouse, whilst less than a foot in height, showed no sensitiveness in their leaves, and had no power of climbing. But when *N. laevis* had grown to a height of 16 inches, there were signs of these powers. The young leaves when first formed stand upright, but soon become inclined; at this period they terminate in a stalk or filament, with the pitcher at the extremity hardly at all developed. The leaves now exhibited slight spontaneous movements; and when the terminal filaments came into contact with a stick, they slowly bent round and firmly seized it. But owing to the subsequent growth of the leaf, this filament became after a time quite slack, though still remaining firmly coiled round the stick. Hence it would appear that the / chief use of the coiling, at least whilst the plant is young, is to support the pitcher with its load of secreted fluid.

SUMMARY ON LEAF-CLIMBERS

Plants belonging to eight families are known to have clasping petioles, and plants belonging to four families climb by the tips of their leaves. In all the species observed by me, with one exception, the young internodes revolve more or less regularly, in some cases as regularly as those of a twining plant. They revolve at various rates, in most cases rather rapidly. Some few can ascend by spirally twining round a support. Differently from most twiners, there is a strong tendency in the same shoot to revolve first in one and then in an opposite direction. The object gained by the revolving movement is to bring the petioles or the tips of the leaves into contact with surrounding objects; and without this aid the plant would be much less successful in climbing. With rare exceptions, the petioles are sensitive only whilst young. They are sensitive on all sides, but in different degrees in different plants; and in some species of clematis the several parts of the same petiole differ much in sensitiveness. The hooked tips of the leaves of the Gloriosa are sensitive only on their inner or inferior surfaces. The petioles are sensitive to a touch and to excessively slight continued pressure, even from a loop of soft thread weighing only the one-sixteenth of a grain (4·05 mg); and there is reason to believe that the rather thick and / stiff petioles of Clematis flammula are sensitive to even much less weight if spread over a wide surface. The petioles always bend towards the side which is pressed or touched, at different rates in different species, sometimes within a few minutes, but generally after a much longer period. After temporary contact with any object, the petiole continues to bend for a considerable time; afterwards it slowly becomes straight again, and can then react. A petiole excited by an extremely slight weight sometimes bends a little, and then becomes accustomed to the stimulus, and either bends no more or becomes straight again, the weight still remaining suspended. Petioles which

53

have clasped an object for some little time cannot recover their original position. After remaining clasped for two or three days, they generally increase much in thickness either throughout their whole diameter or on one side alone; they subsequently become stronger and more woody, sometimes to a wonderful degree; and in some cases they acquire an internal structure like that of the stem or axis.

The young internodes of the Lophospermum as well as the petioles are sensitive to a touch, and by their combined movement seize an object. The flower-peduncles of the *Maurandia semperflorens* revolve spontaneously and are sensitive to a touch, yet are not used for climbing. The leaves of at least two, and probably of most, of the species of clematis, of Fumaria and Adlumia, spontaneously curve from side to side, like the internodes, and are thus better adapted to / seize distant objects. The petioles of the perfect leaves of *Tropaeolum tricolorum*, as well as the tendril-like filaments of the plants whilst young, ultimately move towards the stem or the supporting stick, which they then clasp. These petioles and filaments also show some tendency to contract spirally. The tips of the uncaught leaves of the Gloriosa, as they grow old, contract into a flat spire or helix. These several facts are interesting in relation to true tendrils.

With leaf climbers, as with twining plants, the first internodes which rise from the ground do not, at least in the cases observed by me, spontaneously revolve; nor are the petioles or tips of the first-formed leaves sensitive. In certain species of clematis, the large size of the leaves, together with their habit of revolving, and the extreme sensitiveness of their petioles, appear to render the revolving movement of the internodes superfluous; and this latter power has consequently become much enfeebled. In certain species of Tropaeolum, both the spontaneous movements of the internodes and the sensitiveness of the petioles have become much enfeebled, and in one species have been completely lost. /

CHAPTER III

TENDRIL-BEARERS

Nature of tendrils – Bignoniaceae, various species of, and their different modes of climbing – Tendrils which avoid the light, and creep into crevices – Development of adhesive discs – Excellent adaptations for seizing different kinds of supports – Polemoniaceae – *Cobaea scandens*, much branched and hooked tendrils, their manner of action – Leguminosae – Compositae – Smilaceae – *Smilax aspera*, its inefficient tendrils – Fumariaceae – *Corydalis claviculata*, its state intermediate between that of a leaf-climber and a tendril-bearer

By tendrils I mean filamentary organs, sensitive to contact and used exclusively for climbing. By this definition, spines, hooks, and rootlets, all of which are used for climbing, are excluded. True tendrils are formed by the modification of leaves with their petioles, of flower-peduncles, branches,[1] and perhaps stipules. / Mohl, who includes under the name of tendrils various organs having a similar external appearance, classes them according to their homological nature, as being modified leaves, flower-peduncles, etc. This would be an excellent scheme; but I observe that botanists are by no means

[1] Never having had the opportunity of examining tendrils produced by the modification of branches, I spoke doubtfully about them in this essay when originally published. But since then Fritz Müller has described (*Journal of Linn. Soc.*, vol. ix, p. 344) many striking cases in South Brazil. In speaking of plants which climb by the aid of their branches, more or less modified, he states that the following stages of development can be traced: (1) Plants supporting themselves simply by their branches stretched out at right angles – for example, Chiococca. (2) Plants clasping a support with their unmodified branches, as with Securidaca. (3) Plants climbing by the extremities of their branches which appear like tendrils, as is the case according to Endlicher with Helinus. (4) Plants with their branches much modified and temporarily converted into tendrils, but which may be again / transformed into branches, as with certain Papilionaceous plants. (5) Plants with their branches forming true tendrils, and used exclusively for climbing – as with Strychnos and Caulotretus. Even the unmodified branches become much thickened when they wind round a support. I may add that Mr Thwaites sent me from Ceylon a specimen of an Acacia which had climbed up the trunk of a rather large tree, by the aid of tendril-like, curved or convoluted branchlets, arrested in their growth and furnished with sharp recurved hooks.

unanimous on the homological nature of certain tendrils. Consequently I will describe tendril-bearing plants by natural families,
following Lindley's classification; and this will in most cases keep those
of the same nature together. The species to be described belong to ten
families, and will be given in the following order: Bignoniaceae,
Polemoniaceae, Leguminosae, Compositae, Smilaceae, Fumariaceae.
Cucurbitaceae, Vitaceae, Sapindaceae, Passifloraceae.² /

BIGNONIACEAE

This family contains many tendril-bearers, some twiners, and some
root-climbers. The tendrils always consist of modified leaves. Nine
species of Bignonia, selected by hazard, are here described in order to

Fig. 5   Bignonia
Unnamed species from Kew

show what diversity of structure and action there may be within the
same genus, and to show what remarkable powers some tendrils
possess. The species, taken together, afford connecting links between
twiners, leaf-climbers, tendril-bearers, and root-climbers.

² As far as I can make out, the history of our knowledge of tendrils is as follows:
we have seen that Palm and von Mohl observed about the same time the singular
phenomenon of the spontaneous revolving movement of twining plants. Palm (p.
58), I presume, observed likewise the revolving movement of tendrils; but I do not
feel sure of this, for he says very little on the subject. Dutrochet fully described this
movement of the tendril in the common pea. Mohl first discovered that tendrils are
sensitive to contact; but from some cause, probably from observing too old tendrils,
he was not aware how sensitive they were, and thought that prolonged pressure was
necessary to excite their movement. Professor Asa Gray, in a paper already quoted,
first noticed the extreme sensitiveness and rapidity of the movements of the tendrils
of certain cucurbitaceous plants.

Bignonia (an unnamed species from Kew, closely allied to *B. unguis*, but with smaller and rather broader leaves). A young shoot from a cut-down plant made three revolutions against the sun, at an average rate of 2 hrs 6 m. The stem is thin and flexible; it twined round a slender vertical stick, ascending from left to right, as perfectly and as regularly as any true twining plant. When thus ascending, it makes no use of its tendrils or petioles; but when it twined round a / rather thick stick, and its petioles were brought into contact with it, these curved round the stick, showing that they have some degree of irritability. The petioles also exhibit a slight degree of spontaneous movement; for in one case they certainly described minute, irregular, vertical ellipses. The tendrils apparently curve themselves spontaneously to the same side with the petioles; but from various causes, it was difficult to observe the movement of either the tendrils or petioles, in this and the two following species. The tendrils are so closely similar in all respects to those of *B. unguis*, that one description will suffice.

Bignonia unguis. The young shoots revolve, but less regularly and less quickly than those of the last species. The stem twines imperfectly round a vertical stick, sometimes reversing its direction, in the same manner as described in so many leaf-climbers; and this plant though possessing tendrils, climbs to a certain extent like a leaf-climber. Each leaf consists of a petiole bearing a pair of leaflets, and terminates in a tendril, which is formed by the modification of three leaflets, and closely resembles that above figured (fig. 5). But it is a little larger, and in a young plant was about half an inch in length. It is curiously like the leg and foot of a small bird, with the hind toe cut off. The straight leg or tarsus is longer than the three toes, which are of equal length, and diverging, lie in the same plane. The toes terminate in sharp, hard claws, much curved downwards, like those on a bird's foot. The petiole of the leaf is sensitive to contact; / even a small loop of thread suspended for two days caused it to bend upwards; but the sub-petioles of the two lateral leaflets are not sensitive. The whole tendril, namely, the tarsus and the three toes, are likewise sensitive to contact, especially on their under surfaces. When a shoot grows in the midst of thin branches, the tendrils are soon brought by the revolving movement of the internodes into contact with them; and then one toe of the tendril or more, commonly all three, bend, and after several hours seize fast hold of the twigs, like a bird when perched. If the tarsus of the tendril comes into contact with a twig, it goes on slowly

57

bending, until the whole foot is carried quite round, and the toes pass on each side of the tarsus and seize it. In like manner, if the petiole comes into contact with a twig, it bends round, carrying the tendril, which then seizes its own petiole or that of the opposite leaf. The petioles move spontaneously, and thus, when a shoot attempts to twine round an upright stick, those on both sides after a time come into contact with it, and are excited to bend. Ultimately the two petioles clasp the stick in opposite directions, and the foot-like tendrils, seizing on each other or on their own petioles, fasten the stem to the support with surprising security. The tendrils are thus brought into action, if the stem twines round a thin vertical stick; and in this respect the present species differs from the last. Both species use their tendrils in the same manner when passing through a thicket. This plant is one of the most efficient climbers / which I have observed; and it probably could ascend a polished stem incessantly tossed by heavy storms. To show how important vigorous health is for the action of all the parts, I may mention that when I first examined a plant which was growing moderately well, though not vigorously, I concluded that the tendrils acted only like the hooks on a bramble, and that it was the most feeble and inefficient of all climbers!

Bignonia Tweedyana. This species is closely allied to the last, and behaves in the same manner; but perhaps twines rather better round a vertical stick. On the same plant, one branch twined in one direction and another in an opposite direction. The internodes in one case made two circles, each in 2 hrs 33 m. I was enabled to observe the spontaneous movements of the petioles better in this than in the two preceding species: one petiole described three small vertical ellipses in the course of 11 hrs, whilst another moved in an irregular spire. Some little time after a stem has twined round an upright stick, and is securely fastened to it by the clasping petioles and tendrils, it emits aerial roots from the bases of its leaves; and these roots curve partly round and adhere to the stick. This species of Bignonia, therefore, combines four different methods of climbing generally characteristic of distinct plants, namely, twining, leaf-climbing, tendril-climbing, and root-climbing.

In the three foregoing species, when the foot-like tendril has caught an object, it continues to grow / and thicken, and ultimately becomes wonderfully strong, in the same manner as the petioles of leaf-climbers. If the tendril catches nothing, it first slowly bends downwards, and then its power of clasping is lost. Very soon afterwards it disarticulates itself from the petiole, and drops off like a leaf in autumn. I have seen this process of disarticulation in no other tendrils, for these, when they fail to catch an object, merely wither away.

Bignonia venusta. The tendrils differ considerably from those of the previous species. The lower part, or tarsus, is four times as long as the three toes; these are of equal length and diverge equally, but do not lie in the same plane; their tips are bluntly hooked, and the whole tendril makes an excellent grapnel. The tarsus is sensitive on all sides; but the three toes are sensitive only on their outer surfaces. The sensitiveness is not much developed; for a slight rubbing with a twig did not cause the tarsus or the toes to become curved until an hour had elapsed, and then only in a slight degree. Subsequently they straightened themselves. Both the tarsus and toes can seize well hold of sticks. If the stem is secured, the tendrils are seen spontaneously to sweep large ellipses; the two opposite tendrils moving independently of one another. I have no doubt, from the analogy of the two following allied species, that the petioles also move spontaneously; but they are not irritable like those of *B. unguis* and *B. Tweedyana*. The young internodes sweep large circles, one being completed in 2 hrs 15 m, and / a second in 2 hrs 55 m. By these combined movements of the internodes, petioles, and grapnel-like tendrils, the latter are soon brought into contact with surrounding objects. When a shoot stands near an upright stick, it twines regularly and spirally round it. As it ascends, it seizes the stick with one of its tendrils, and, if the stick be thin, the right- and left-hand tendrils are alternately used. This alternation follows from the stem necessarily taking one twist round its own axis for each completed circle.

The tendrils contract spirally a short time after catching any object; those which catch nothing merely bend slowly downwards. But the whole subject of the spiral contraction of tendrils will be discussed after all the tendril-bearing species have been described.

Bignonia littoralis. The young internodes revolve in large ellipses. An internode bearing immature tendrils made two revolutions, each in 3 hrs 50 m; but when grown older with the tendrils mature, it made two ellipses, each at the rate of 2 hrs 44 m. This species, unlike the preceding, is incapable of twining round a stick: this does not appear to be due to any want of flexibility in the internodes or to the action of the tendrils, and certainly not to any want of the revolving power; nor can I account for the fact. Nevertheless the plant readily ascends a thin upright stick by seizing a point above with its two opposite tendrils, which then contract spirally. If the tendrils seize nothing, they do not become spiral. / The species last described, ascended a vertical stick by twining spirally and by seizing it alternately with its opposite tendrils, like a sailor pulling himself up a rope, hand over hand; the present species pulls itself up, like a sailor seizing with both hands together a rope above his head.

The tendrils are similar in structure to those of the last species. They continue growing for some time, even after they have clasped an object. When fully grown, though borne by a young plant, they are nine inches in length. The three divergent toes are shorter relatively to the tarsus than in the former species; they are blunt at their tips and but slightly hooked; they are not quite equal in length, the middle one being rather longer than the others. Their outer surfaces are highly sensitive; for when lightly rubbed with a twig, they became perceptibly , curved in 4 m and greatly curved in 7 m. In 7 hrs they became straight again and were ready to react. The tarsus, for the space of one inch close to the toes, is sensitive, but in a rather less degree than the toes; for the latter, after a slight rubbing, became curved in about half the time. Even the middle part of the tarsus is sensitive to prolonged contact, as soon as the tendril has arrived at maturity. After it has grown old, the sensitiveness is confined to the toes, and these are only able to curl very slowly round a stick. A tendril is perfectly ready to act, as soon as the three toes have diverged, and at this period their outer surfaces first become irritable. The irritability spreads but little from one part when / excited to another: thus, when a stick was caught by the part immediately beneath the three toes, these seldom clasped it, but remained sticking straight out.

The tendrils revolve spontaneously. The movement begins before the tendril is converted into a three-pronged grapnel by the divergence of the toes, and before any part has become sensitive; so that the revolving movement is useless at this early period. The movement is, also, now slow, two ellipses being completed conjointly in 24 hrs 18 m. A mature tendril made an ellipse in 6 hrs; so that it moved much more slowly than the internodes. The ellipses which were swept, both in a vertical and horizontal plane, were of large size. The petioles are not in the least sensitive, but revolve like the tendrils. We thus see that the young internodes, the petioles, and the tendrils all continue revolving together, but at different rates. The movement of the tendrils which rise opposite one another are quite independent. Hence, when the whole shoot is allowed freely to revolve, nothing can be more intricate than the course followed by the extremity of each tendril. A wide space is thus irregularly searched for some object to be grasped.

One other curious point remains to be mentioned. In the course of a few days after the toes have closely clasped a stick, their blunt extremities become developed, though not invariably, into irregular disc-like balls which have the power of adhering firmly to the wood. As similar cellular outgrowths will be / fully described under *B. capreolata*, I will here say nothing more about them.

Bignonia aequinoctialis, var. *Chamberlaynii*. The internodes, the elongated non-sensitive petioles, and the tendrils all revolve. The stem does not twine, but ascends a vertical stick in the same manner as the last species. The tendrils also resemble those of the last species, but are shorter; the three toes are more unequal in length, the two outer ones being about one-third shorter and rather thinner than the middle toe; but they vary in this respect. They terminate in small hard points; and what is important, cellular adhesive discs are not developed. The reduced size of two of the toes as well as their lessened sensitiveness, seem to indicate a tendency to abortion; and on one of my plants the first-formed tendrils were sometimes simple, that is, were not divided into three toes. We are thus naturally led to the three following species with undivided tendrils:

Bignonia speciosa. The young shoots revolve irregularly, making narrow ellipses, spires, or circles, at rates varying from 3 hrs 30 m to 4 hrs 40 m; but they show no tendency to twine. Whilst the plant is young and does not require a support, tendrils are not developed. Those borne by a moderately young plant were five inches in length. They revolve spontaneously, as do the short and non-sensitive petioles. When rubbed, they slowly bend to the rubbed side and subsequently straighten themselves; but they are not highly sensitive. There is something strange in / their behaviour: I repeatedly placed close to them, thick and thin, rough and smooth sticks and posts, as well as string suspended vertically, but none of these objects were well seized. After clasping an upright stick, they repeatedly loosed it again, and often would not seize it at all, or their extremities did not coil closely round. I have observed hundreds of tendrils belonging to various Cucurbitaceous, Passifloraceous, and Leguminous plants, and never saw one behave in this manner. When, however, my plant had grown to a height of eight or nine feet, the tendrils acted much better. They now seized a thin, upright stick horizontally, that is, at a point on their own level, and not some way up the stick as in the case of all the previous species. Nevertheless, the non-twining stem was enabled by this means to ascend the stick.

The extremity of the tendril is almost straight and sharp. The whole terminal portion exhibits a singular habit, which in an animal would be called an instinct; for it continually searches for any little crevice or hole into which to insert itself. I had two young plants; and, after having observed this habit, I placed near them posts, which had been bored by beetles, or had become fissured by drying. The tendrils, by their own movement and by that of the internodes, slowly travelled over the surface of the wood, and when the apex came to a hole or fissure it inserted itself; in order to effect this the extremity for a length of half or quarter of an inch, would often bend itself at right angles to the basal part. I have watched this process / between twenty and thirty times. The same tendril would frequently withdraw from one hole and insert its point into a second hole. I have also seen a tendril keep its point, in one case for 20 hrs and in another for 36 hrs, in a minute hole, and then withdraw it. Whilst the point is thus temporarily inserted, the opposite tendril goes on revolving.

The whole length of a tendril often fits itself closely to any surface of wood with which it has come into contact; and I have observed one bent at right angles, from having entered a wide and deep fissure, with its apex abruptly re-bent and inserted into a minute lateral hole. After a tendril has clasped a stick, it contracts spirally; if it remains unattached it hangs straight downwards. If it has merely adapted itself to the inequalities of a thick post, though it has clasped nothing, or if it has inserted its apex into some little fissure, this stimulus suffices to induce spiral contraction; but the contraction always draws the tendril away from the post. So that in every case these movements, which seem so nicely adapted for some purpose, were useless. On one occasion, however, the tip became permanently jammed into a narrow fissure. I fully expected, from the analogy of B. capreolata and B. littoralis, that the tips would have been developed into adhesive discs; but I could never detect even a trace of this process. There is therefore at present something unintelligible about the habits of this plant.

Bignonia picta. This species closely resembles the / last in the structure and movements of its tendrils. I also casually examined a fine growing plant of the allied *B. Lindleyi*, and this apparently behaved in all respects in the same manner.

Bignonia capreolata. We now come to a species having tendrils of a different type; but first for the internodes. A young shoot made three large revolutions, following the sun, at an average rate of 2 hrs 23 m. The stem is thin and flexible, and I have seen one make four regular spiral turns round a thin upright stick, ascending of course from right to left, and therefore in a reversed direction compared with the before described species. Afterwards, from the interference of the tendrils, it ascended either straight up the stick or in an irregular spire. The tendrils are in some respects highly remarkable. In a young plant they were about 2½ inches in length and much branched, the five chief branches apparently representing two pairs of leaflets and a terminal one. Each branch is, however, bifid or more commonly trifid towards the extremity, with the points blunt yet distinctly hooked. A tendril bends to any side which is lightly rubbed, and subsequently becomes straight again; but a loop of thread weighing ¼th of a grain produced no effect. On two occasions the terminal branches became slightly curved in 10 m after they had touched a stick; and in 30 m the tips were curled quite round it. The basal part is less sensitive. The tendrils revolved in an apparently capricous manner, sometimes very slightly or not at all; at other times they / described large regular ellipses. I could detect no spontaneous movement in the petioles of the leaves.

Whilst the tendrils are revolving more or less regularly, another remarkable movement takes place, namely, a slow inclination from the light towards the darkest side of the house. I repeatedly changed the position of my plants, and some little time after the revolving movement had ceased, the successively formed tendrils always ended by pointing to the darkest side. When I placed a thick post near a tendril, between it and the light, the tendril pointed in that direction. In two instances a pair of leaves stood so that one of the two tendrils was directed towards the light and the other to the darkest side of the house; the latter did not move, but the opposite one bent itself first upwards and then right over its fellow, so that the two became parallel, one above the other, both pointing to the dark: I then turned the plant half round; and the tendril which had turned over recovered its original position, and the opposite one which had not before moved, now turned over to the dark side. Lastly, on another plant, three pairs of tendrils were produced at the same time by three shoots, and all happened to be differently directed: I placed the pot in a box open only on one side, and obliquely facing the light; in two days all six tendrils pointed with unerring truth to the darkest corner of the box, though to do this each had to bend in a different manner. Six wind-vanes could not have more truly shown the direction of the wind, than did / these branched tendrils the course of the stream of light which entered the box. I left these tendrils undisturbed for above 24 hrs, and then turned the pot half round; but they had now lost their power of movement, and could not any longer avoid the light.

When a tendril has not succeeded in clasping a support, either through its own revolving movement or that of the shoot, or by turning towards any object which intercepts the light, it bends vertically downwards and then towards its own stem, which it seizes together with the supporting stick, if there be one. A little aid is thus given in keeping the stem secure. If the tendril seizes nothing, it does not contract spirally, but soon withers away and drops off. If it seizes an object, all the branches contract spirally.

I have stated that after a tendril has come into contact with a stick, it bends round it in about half an hour; but I repeatedly observed, as in the case of *B. speciosa* and its allies, that it often again loosed the stick; sometimes seizing and loosing the same stick three or four times. Knowing that the tendrils avoided the light, I gave them a glass tube blackened within, and a well-blackened zinc plate: the branches curled round the tube and abruptly bent themselves round the edges of the zinc plate; but they soon recoiled from these objects with what I can only call disgust, and straightened themselves. I then placed a post with extremely rugged bark close to a pair of tendrils; twice they touched it for an hour or two, and twice they withdrew; at last one of the hooked extremities / curled round and firmly seized an excessively minute projecting point of bark, and then the other branches spread themselves out, following with accuracy every inequality of the surface. I afterwards placed near the plant a post without bark but much fissured, and the points of the tendrils crawled into all the crevices in a beautiful manner. To my surprise, I observed that the tips of the immature tendrils, with the branches not yet fully separated, likewise crawled just like roots into the minutest crevices. In two or three days after the tips had thus crawled into the crevices, or after their hooked ends had seized minute points, the final process, now to be described, commenced.

This process I discovered by having accidentally left a piece of wool near a tendril; and this led me to bind a quantity of flax, moss, and wool loosely round sticks, and to place them near tendrils. The wool must not be dyed, for these tendrils are excessively sensitive to some poisons. The hooked points soon caught hold of the fibres, even loosely floating fibres, and now there was no recoiling; on the contrary, the excitement caused the hooks to penetrate the fibrous mass and to curl inwards, so that each hook caught firmly one or two fibres, or a small bundle of them. The tips and the inner surfaces of the hooks now began to swell, and in two or three days were visibly enlarged. After a few more days the hooks were converted into whitish, irregular balls, rather above the $\frac{1}{20}$th of an inch (1·27 mm) in diameter, formed of coarse cellular tissue, / which sometimes wholly enveloped and concealed the hooks themselves. The surfaces of these balls secrete some viscid resinous matter, to which the fibres of the flax, etc., adhere. When a fibre has become fastened to the surface, the cellular tissue does not grow directly beneath it, but continues to grow closely on each side; so that when several adjoining fibres, though excessively thin, were caught, so many crests of cellular matter, each not as thick as a human hair, grew up between them, and these, arching over on both sides, adhered firmly together. As the whole surface of the ball continues to grow, fresh fibres adhere and are afterwards enveloped; so that I have seen a little ball with between fifty and sixty fibres of flax crossing it at various angles and all embedded more or less deeply. Every gradation in the process could be followed – some fibres merely sticking to the surface, others lying in more or less deep furrows, or deeply embedded, or passing through the very centre of the cellular ball. The embedded fibres are so closely clasped that they cannot be withdrawn. The outgrowing tissue has so strong a tendency to unite, that two balls produced by distinct tendrils sometimes unite and grow into a single one.

On one occasion, when a tendril had curled round a stick, half an inch in diameter, an adhesive disc was formed; but this does not generally occur in the case of smooth sticks or posts. If, however, the tip catches a minute projecting point, the other branches form discs, especially if they find crevices to crawl / into. The tendrils failed to attach themselves to a brick wall.

I infer from the adherence of the fibres to the discs or balls, that these secrete some resinous adhesive matter; and more especially from such fibres becoming loose if immersed in sulphuric ether. This fluid likewise removes small, brown, glistening points which can generally be seen on the surfaces of the older discs. If the hooked extremities of the tendrils do not touch anything, discs, as far as I have seen, are never formed;[3] but temporary contact during a moderate time suffices to cause their development. I have seen eight discs formed on the same tendril. After their development the tendrils contract spirally, and become woody and very strong. A tendril in this state supported nearly seven ounces, and would apparently have supported a considerably greater weight, had not the fibres of flax to which the discs were attached yielded.

From the facts now given, we may infer that though the tendrils of this Bignonia can occasionally adhere to smooth cylindrical sticks and often to rugged bark, yet that they are specially adapted to climb trees clothed with lichens, mosses, or other such productions; and I hear from Professor Asa Gray that the *Polypodium incanum* abounds on the forest trees in the districts of / North America where this species of Bignonia grows. Finally, I may remark how singular a fact it is that a leaf should be metamorphosed into a branched organ which turns from the light, and which can by its extremities either crawl like roots into crevices, or seize hold of minute projecting points, these extremities afterwards forming cellular outgrowths which secrete an adhesive cement, and then envelop by their continued growth the finest fibres.

Eccremocarpus scaber (Bignoniaceae). Plants, though growing pretty well in my greenhouse, showed no spontaneous movements in their shoots or tendrils; but when removed to the hothouse, the young internodes revolved at rates varying from 3 hrs 15 m to 1 hr 13 m. One large circle was swept at this latter unusually quick rate; but generally the circles or ellipses were small, and sometimes the course pursued was quite irregular. An internode, after making several revolutions, sometimes stood still for 12 hrs or 18 hrs, and then recommenced revolving. Such strongly marked interruptions in the movements of the internodes I have observed in hardly any other plant.

The leaves bear four leaflets, themselves subdivided, and terminate in much-branched tendrils. The main petiole of the leaf, whilst young,

---

[3] Fritz Müller states (ibid., p. 348) that in south Brazil the trifid tendrils of Haplolophium (one of the Bignoniaceae), without having come into contact with any object, terminate in smooth shining discs. These, however, after adhering to any object, sometimes become considerably enlarged.

moves spontaneously, and follows nearly the same irregular course and at about the same rate as the internodes. The movement to and from the stem is the most conspicuous, and I have seen the chord of a curved petiole which formed an angle of 59° with the stem, in an / hour afterwards making an angle of 106°. The two opposite petioles do not move together, and one is sometimes so much raised as to stand close to the stem, whilst the other is not far from horizontal. The basal part of the petiole moves less than the distal part. The tendrils, besides being carried by the moving petioles and internodes, themselves move spontaneously; and the opposite tendrils occasionally move in opposite directions. By these combined movements of the young internodes, petioles, and tendrils, a considerable space is swept in search of a support.

In young plants the tendrils are about three inches in length: they bear two lateral and two terminal branches; and each branch bifurcates twice, with the tips terminating in blunt double hooks, having both points directed to the same side. All the branches are sensitive on all sides; and after being lightly rubbed, or after coming into contact with a stick, bend in about 10 m. One which had become curved in 10 m after a light rub, continued bending for between 3 hrs and 4 hrs, and became straight again in 8 hrs or 9 hrs. Tendrils, which have caught nothing, ultimately contract into an irregular spire, as they likewise do, only much more quickly, after clasping a support. In both cases the main petiole bearing the leaflets, which is at first straight and inclined a little upwards, moves downwards, with the middle part bent abruptly into a right angle; but this is seen in *E. miniatus* more plainly than in *E. scaber*. The tendrils in this genus act in some respects like those of *Bignonia / capreolata*; but the whole does not move from the light, nor do the hooked tips become enlarged into cellular discs. After the tendrils have come into contact with a moderately thick cylindrical stick or with rugged bark, the several branches may be seen slowly to lift themselves up, change their positions, and again come into contact with the supporting surface. The object of these movements is to bring the double hooks at the extremities of the branches, which naturally face in all directions, into contact with the wood. I have watched a tendril, half of which had bent itself at right angles round the sharp corner of a square post, neatly bring every single hook into contact with both rectangular surfaces. The appearance suggested the belief, that though the whole tendril is not sensitive to light, yet that the tips are so, and that they turn and twist themselves

67

towards any dark surface. Ultimately the branches arrange themselves very neatly to all the irregularities of the most rugged bark, so that they resemble in their irregular course a river with its branches, as engraved on a map. But when a tendril has wound round a rather thick stick, the subsequent spiral contraction generally draws it away and spoils the neat arrangement. So it is, but not in quite so marked a manner, when a tendril has spread itself over a large, nearly flat surface of rugged bark. We may therefore conclude that these tendrils are not perfectly adapted to seize moderately thick sticks or rugged bark. If a thin stick or twig is placed near a tendril, the terminal branches wind quite round it, / and then seize their own lower branches or the main stem. The stick is thus firmly, but not neatly, grasped. What the tendrils are really adapted for, appears to be such objects as the thin culms of certain grasses, or the long flexible bristles of a brush, or thin rigid leaves such as those of the asparagus, all of which they seize in an admirable manner. This is due to the extremities of the branches close to the little hooks being extremely sensitive to a touch from the thinnest object, which they consequently curl round and clasp. When a small brush, for instance, was placed near a tendril, the tips of each sub-branch seized one, two, or three of the bristles; and then the spiral contraction of the several branches brought all these little parcels close together, so that thirty or forty bristles were drawn into a single bundle, which afforded an excellent support.

POLEMONIACEAE

Cobaea scandens. This is an excellently constructed climber. The tendrils on a fine plant were eleven inches long, with the petiole bearing two pairs of leaflets, only two and a half inches in length. They revolve more rapidly and vigorously than those of any other tendril-bearer observed by me, with the exception of one kind of *Passiflora*. Three large, nearly circular sweeps, directed against the sun were completed, each in 1 hr 15 m; and two other circles in 1 hr 20 m and 1 hr 23 m. Sometimes a tendril travels in a much inclined position, and sometimes nearly upright. The lower part moves but little and the petiole not at all; nor do / the internodes revolve; so that here we have the tendril alone moving. On the other hand, with most of the species of *Bignonia* and the *Eccremocarpus*, the internodes, tendrils, and petioles all revolved. The long, straight, tapering main stem of the tendril of the *Cobaea* bears alternate branches; and each branch is several times divided, with the finer branches as thin as very thin bristles and extremely flexible, so that they are blown about by a breath of air; yet they are strong and highly elastic. The extremity of each branch is a little flattened, and terminates in a minute double (though sometimes single) hook, formed of a hard, translucent, woody substance, and as sharp as the finest needle. On a tendril which was eleven inches long I counted ninety-four of these beautifully constructed little hooks. They readily catch soft wood, or gloves, or the skin of the naked hand. With the exception of these hardened hooks, and of the basal part of the central stem, every part of every branchlet is highly sensitive on all sides to a slight touch, and bends in a few minutes towards the touched side. By lightly rubbing several sub-branches on opposite sides, the whole tendril rapidly assumed an extraordinarily crooked shape. These movements from contact do not interfere with the ordinary revolving movement. The branches, after becoming greatly curved from being touched, straighten themselves at a quicker rate than in almost any other tendril seen by me, namely in between half an hour and an hour. After the tendril has caught any object, spiral contraction likewise / begins after an unusually short interval of time, namely, in about twelve hours.

Before the tendril is mature, the terminal branchlets cohere, and the hooks are curled closely inwards. At this period no part is sensitive to a touch; but as soon as the branches diverge and the hooks stand out, full sensitiveness is acquired. It is a singular circumstance that immature tendrils revolve at their full velocity before they become sensitive, but in a useless manner, as in this state they can catch nothing. This want of perfect co-adaptation, though only for a short time, between the structure and the functions of a climbing plant is a rare event. A tendril, as soon as it is ready to act, stands, together with the supporting petiole, vertically upwards. The leaflets borne by the petiole are at this time quite small, and the extremity of the growing stem is bent to one side so as to be out of the way of the revolving tendril, which sweeps large circles directly overhead. The tendrils thus revolve in a position well adapted for catching objects standing above; and by this means the ascent of the plant is favoured. If no object is caught, the leaf with its tendril bends downwards and ultimately assumes a horizontal position. An open space is thus left for the next succeeding and younger tendril to stand vertically upwards and to revolve freely. As soon as an old tendril bends downwards, it loses all power of movement, and contracts spirally into an entangled mass. Although the tendrils revolve with unusual rapidity, the movement lasts for only a short / time. In a plant placed in the hothouse and growing vigorously, a tendril revolved for not longer than 36 hours, counting from the period when it first became sensitive; but during this period it probably made at least 27 revolutions.

When a revolving tendril strikes against a stick, the branches quickly bend round and clasp it. The little hooks here play an important part, as they prevent the branches from being dragged away by the rapid revolving movement, before they have had time to clasp the stick securely. This is especially the case when only the extremity of a branch has caught hold of a support. As soon as a tendril has bent round a smooth stick or a thick rugged post, or has come into contact with planed wood (for it can adhere temporarily even to so smooth a surface as this), the same peculiar movements may be observed as those described under *Bignonia capreolata* and *Eccremocarpus*. The branches repeatedly lift themselves up and down; those which have their hooks already directed downwards remaining in this position and securing the tendril, whilst the others twist about until they succeed in arranging themselves in conformity with every irregularity of the surface, and in bringing their hooks into contact with the wood. The use of the hooks was well shown by giving the tendrils tubes and slips of glass to catch; for these, though temporarily seized, were invariably lost, either during the rearrangement of the branches or ultimately when spiral contraction ensued. /

The perfect manner in which the branches arranged themselves, creeping like rootlets over every inequality of the surface and into any deep crevice, is a pretty sight; for it is perhaps more effectually performed by this than by any other species. The action is certainly more conspicuous, as the upper surfaces of the main stem, as well as of every branch to the extreme hooks, are angular and green, whilst the lower surfaces are rounded and purple. I was led to infer, as in former cases, that a less amount of light guided these movements of the branches of the tendrils. I made many trials with black and white cards and glass tubes to prove it, but failed from various causes; yet these trials countenanced the belief. As a tendril consists of a leaf split into numerous segments, there is nothing surprising in all the segments turning their upper surfaces towards the light, as soon as the tendril is caught and the revolving movement is arrested. But this will not account for the whole movement, for the segments actually bend or curve to the dark side besides turning round on their axes so that their upper surfaces may face the light.

When the *Cobaea* grows in the open air, the wind must aid the extremely flexible tendrils in seizing a support, for I found that a mere breath sufficed to cause the extreme branches to catch hold by their hooks of twigs, which they could not have reached by the revolving movement. It might have been thought that a tendril, thus hooked by the extremity of a single branch, could not have fairly grasped its support. / But several times I watched cases like the following: a tendril caught a thin stick by the hooks of one of its two extreme branches; though thus held by the tip, it still tried to revolve, bowing itself to all sides, and by this movement the other extreme branch soon caught the stick. The first branch then loosed itself, and, arranging its hooks, again caught hold. After a time, from the continued movement of the tendril, the hooks of a third branch caught hold. No other branches, as the tendril then stood, could possibly have touched the stick. But before long the upper part of the main stem began to contract into an open spire. It thus dragged the shoot which bore the tendril towards the stick; and as the tendril continually tried to revolve, a fourth branch was brought into contact. And lastly, from the spiral contraction travelling down both the main stem and the branches, all of them, one after another, were ultimately brought into contact with the stick. They then wound themselves round it and round one another, until the whole tendril was tied together in an inextricable knot. The tendrils, though at first quite flexible, after having clasped a support for a time, become more rigid and stronger than they were at first. Thus the plant is secured to its support in a perfect manner.

LEGUMINOSAE

Pisum sativum. The common pea was the subject of a valuable memoir by Dutrochet,[4] who discovered that the internodes and tendrils / revolve in ellipses. The ellipses are generally very narrow, but sometimes approach to circles. I several times observed that the longer axis slowly changed its direction, which is of importance, as the tendril thus sweeps a wider space. Owing to this change of direction, and likewise to the movement of the stem towards the light, the successive

[4] *Comptes Rendus*, vol. xvii, 1843, p. 989.

irregular ellipses generally form an irregular spire. I have thought it worthwhile to annex a tracing of the course pursued by the upper internode (the movement of the tendril being neglected) of a young plant from 8.40 a.m. to 9.15 p.m. The course was traced on a hemispherical glass placed over the plant, and the dots with figures give the hours of observation; each dot being joined by a straight line. No doubt all the lines would have been curvilinear if the course had been observed at much shorter intervals. The extremity of the petiole, from which the young tendril arose, was two inches from the glass, so that if a pencil two inches in length could have been affixed to the petiole, it would have traced the annexed figure on the underside of the glass; but it must be remembered that the figure is reduced by one-half. Neglecting the first great sweep towards the light from the figure

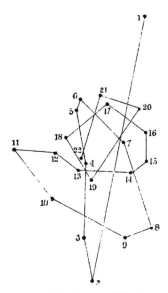

Fig. 6   Side of room with window

Diagram showing the movement of the upper internode of the common pea, traced on a hemispherical glass, and transferred to paper; reduced one-half in size. (1 Aug.)

No.	H M	No.	H M	No.	H M
1	8 46 a.m.	9	1 55 p.m.	16	5 25 p.m.
2	10 0	10	2 25	17	5 50
3	11 0	11	3 0	18	6 25
4	11 37	12	3 30	19	7 0
5	12 7 p.m.	13	3 48	20	7 45
6	12 30	14	4 40	21	8 30
7	1 0	15	5 5	22	9 15
8	1 30				

1 to 2, the end of the petiole swept a space 4 inches across in one direction, and 3 inches in another. As a full-grown tendril is considerably above two inches in length, and as the tendril itself bends and revolves in harmony with the internode, a considerably wider space is swept than is here represented on a reduced scale. Dutrochet / observed the completion of an ellipse in 1 hr 20 m; and I saw one completed in 1 hr 30 m. The direction followed is variable, either with or against the sun.

Dutrochet asserts that the petioles of the leaves spontaneously revolve, as well as the young internodes and tendrils; but he does not say that he / secured the internodes; when this was done, I could never detect any movement in the petiole, except to and from the light.

The tendrils, on the other hand, when the internodes and petioles are secured, describe irregular spires or regular ellipses, exactly like those made by the internodes. A young tendril, only 1⅛ inch in length, revolved. Dutrochet has shown that when a plant is placed in a room, so that the light enters laterally, the internodes travel much quicker to the light than from it: on the other hand, he asserts that the tendril itself moves from the light towards the dark side of the room. With due reference to this great observer, I think he was mistaken, owing to his not having secured the internodes. I took a young plant with highly sensitive tendrils, and tied the petiole so that the tendril alone could move; it completed a perfect ellipse in 1 hr 30 m; I then turned the plant partly round, but this made no change in the direction of the succeeding ellipse. The next day I watched a plant similarly secured until the tendril (which was highly sensitive) made an ellipse in a line exactly to and from the light; the movement was so great that the tendril at the two ends of its elliptical course bent itself a little beneath the horizon, thus travelling more than 180 degrees; but the curvature was fully as great towards the light as towards the dark side of the room. I believe Dutrochet was misled by not having secured the internodes, and by having observed a plant of which the internodes and tendrils no longer / curved in harmony together, owing to inequality of age.

Dutrochet made no observations on the sensitiveness of the tendrils. These, whilst young and about an inch in length with the leaflets on the petiole only partially expanded, are highly sensitive; a single light touch with a twig on the inferior or concave surface near the tip caused them to bend quickly, as did occasionally a loop of thread weighing one-seventh of a grain (9·25 mg). The upper or convex surface is

73

barely or not at all sensitive. Tendrils, after bending from a touch, straighten themselves in about two hours, and are then ready to act again. As soon as they begin to grow old, the extremities of their two or three pairs of branches become hooked, and they then appear to form an excellent grappling instrument; but this is not the case. For at this period they have generally quite lost their sensitiveness; and when hooked on to twigs, some were not at all affected, and others required from 18 hrs to 24 hrs before clasping such twigs; nevertheless, they were able to utilize the last vestige of irritability owing to their extremities being hooked. Ultimately the lateral branches contract spirally, but not the middle or main stem.

Lathyrus aphaca. This plant is destitute of leaves, except during a very early age, these being replaced by tendrils, and the leaves themselves by large stipules. It might therefore have been expected that the tendrils would have been highly organized, but this is not so. They are moderately long, thin, and unbranched, / with their tips slightly curved. Whilst young they are sensitive on all sides, but chiefly on the concave side of the extremity. They have no spontaneous revolving power, but are at first inclined upwards at an angle of about 45°, then move into a horizontal position, and ultimately bend downwards. The young internodes, on the other hand, revolve in ellipses, and carry with them the tendrils. Two ellipses were completed, each in nearly 5 hrs; their longer axes were directed at about an angle of 45° to the axis of the previously made ellipse.

Lathyrus grandiflorus. The plants observed were young and not growing vigorously, yet sufficiently so, I think, for my observations to be trusted. If so, we have the rare case of neither internodes nor tendrils revolving. The tendrils of vigorous plants are above 4 inches in length, and are often twice divided into three branches; the tips are curved and are sensitive on their concave sides; the lower part of the central stem is hardly at all sensitive. Hence this plant appears to climb simply by its tendrils being brought, through the growth of the stem, or more efficiently by the wind, into contact with surrounding objects, which they then clasp. I may add that the tendrils, or the internodes, or both, of *Vicia sativa* revolve.

COMPOSITAE

Mutisia clematis. The immense family of the Compositae is well known to include very few climbing plants. We have seen in the table in the first chapter that *Mikania scandens* is a regular twiner, and F. Müller informs me that in S. / Brazil there is another species which is a leaf-climber. *Mutisia* is the only genus in the family, as far as I can learn, which bears tendrils: it is therefore interesting to find that these, though rather less metamorphosed from their primordial foliar condition than are most other tendrils, yet display all the ordinary characteristic movements, both those that are spontaneous and those which are excited by contact.

The long leaf bears seven or eight alternate leaflets, and terminates in a tendril which, in a plant of considerable size, was 5 inches in length. It consists generally of three branches; and these, although much elongated, evidently represent the petioles and midribs of three leaflets; for they closely resemble the same parts in an ordinary leaf, in being rectangular on the upper surface, furrowed, and edged with green. Moreover, the green edging of the tendrils of young plants sometimes expands into a narrow lamina or blade. Each branch is curved a little downwards, and is slightly hooked at the extremity.

A young upper internode revolved, judging from three revolutions, at an average rate of 1 hr 38 m; it swept ellipses with the longer axes directed at right angles to one another; but the plant, apparently, cannot twine. The petioles and the tendrils are both in constant movement. But their movement is slower and much less regularly elliptical than that of the internodes. They appear to be much affected by the light, for the whole leaf usually sinks down during the / night and rises during the day, moving, also, during the day in a crooked course to the west. The tip of the tendril is highly sensitive on the lower surface; and one which was just touched with a twig became perceptibly curved in 3 m, and another in 5 m; the upper surface is not at all sensitive; the sides are moderately sensitive, so that two branches which were rubbed on their inner sides converged and crossed each other. The petiole of the leaf and the lower parts of the tendril, halfway between the upper leaflet and the lowest branch, are not sensitive. A tendril after curling from a touch became straight again in about 6 hrs, and was ready to react; but one that had been so roughly rubbed as to have coiled into a helix did not become perfectly straight until after 13 hrs. The tendrils retain their sensibility to an unusually late age; for one borne by a leaf with five or six fully developed leaves above, was still active. If a tendril catches nothing, after a considerable interval of time the tips of the branches curl a little inwards; but if it clasps some object, the whole contracts spirally.

SMILACEAE

Smilax aspera, var. *maculata*. Aug. St-Hilaire[5] considers that the tendrils, which rise in pairs from the petiole, are modified lateral leaflets; but Mohl (p. 41) ranks them as modified stipules. These tendrils are from 1½ to 1¾ inches in length, are thin, and have slightly curved, pointed extremities. They diverge a little from each other, and stand at first nearly upright. When lightly rubbed on either / side, they slowly bend to that side, and subsequently become straight again. The back or convex side when placed in contact with a stick became just perceptibly curved in 1 hr 20 m, but did not completely surround it

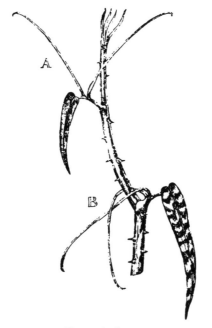

Fig. 7   *Smilax aspera*

until 48 hrs had elapsed; the concave side of another became considerably curved in 2 hrs and clasped a stick in 5 hrs. As the pairs of tendrils grow old, one tendril diverges more and more from the other,

---

[5] *Leçons de Botanique*, 1841, p. 170.

and both slowly bend backwards and downwards, so that after a time they project on the opposite side / of the stem to that from which they arise. They then still retain their sensitiveness, and can clasp a support placed *behind* the stem. Owing to this power, the plant is able to ascend a thin upright stick. Ultimately the two tendrils belonging to the same petiole, if they do not come into contact with any object, loosely cross each other behind the stem, as at B, in fig. 7. This movement of the tendrils towards and round the stem is, to a certain extent, guided by their avoidance of the light; for when a plant stood so that one of the two tendrils was compelled in thus slowly moving to travel towards the light, and the other from the light, the latter always moved, as I repeatedly observed, more quickly than its fellow. The tendrils do not contract spirally in any case. Their chance of finding a support depends on the growth of the plant, on the wind, and on their own slow backward and downward movement, which, as we have just seen, is guided, to a certain extent, by the avoidance of the light; for neither the internodes nor the tendrils have any proper revolving movement. From this latter circumstance, from the slow movements of the tendrils after contact (though their sensitiveness is retained for an unusual length of time), from their simple structure and shortness, this plant is a less perfect climber than any other tendril-bearing species observed by me. The plant whilst young and only a few inches in height, does not produce any tendrils; and considering that it grows to only about 8 feet in height, that the stem is zigzag and is furnished, as well as the petioles, with / spines, it is surprising that it should be provided with tendrils, comparatively inefficient though these are. The plant might have been left, one would have thought, to climb by the aid of its spines alone, like our brambles. As, however, it belongs to a genus, some of the species of which are furnished with much longer tendrils, we may suspect that it possesses these organs solely from being descended from progenitors more highly organized in this respect.

FUMARIACEAE

Corydalis claviculata. According to Mohl (p. 43), the extremities of the branched stem, as well as the leaves, are converted into tendrils. In the specimens examined by me all the tendrils were certainly foliar, and it is hardly credible that the same plant should produce tendrils of a widely different homological nature. Nevertheless, from this statement by

77

Mohl, I have ranked this species among the tendril-bearers; if classed exclusively by its foliar tendrils, it would be doubtful whether it ought not to have been placed among the leaf-climbers, with its allies, *Fumaria* and *Adlumia*. A large majority of its so-called tendrils still bear leaflets, though excessively reduced in size; but some few of them may properly be designated as tendrils, for they are completely destitute of laminae or blades. Consequently, we here behold a plant in an actual state of transition from a leaf-climber to a tendril-bearer. Whilst the plant is rather young, only the outer leaves, but when full-grown all the leaves, have their extremities converted into more or less perfect tendrils. I have examined specimens / from one locality alone, viz. Hampshire; and it is not improbable that plants growing under different conditions might have their leaves a little more or less changed into true tendrils.

Whilst the plant is quite young, the first-formed leaves are not modified in any way, but those next formed have their terminal leaflets reduced in size, and soon all the leaves assume the structure represented in the following drawing. This leaf bore nine leaflets; the lower ones being much subdivided. The terminal portion of the petiole, about 1½ inch in length (above the leaflet *f*), is thinner and more elongated than the lower part, and may be considered as the tendril. The leaflets borne by this part are greatly reduced in size, being, on an average, about the tenth of an inch in length and very narrow; one small leaflet measured one-twelfth of an inch in length and one-seventy-fifth in breadth (2·116 mm. and 0·339 mm.), so that it was almost microscopically minute. All the reduced leaflets have branching nerves, and terminate in little spines, like those of the fully developed leaflets. Every gradation could be traced, until we come to branchlets (as *a* and *d* in the figure) which show no vestige of a lamina or blade. Occasionally all the terminal branchlets of the petiole are in this condition, and we then have a true tendril.

The several terminal branches of the petiole bearing the much reduced leaflets (*a, b, c, d*) are highly sensitive, for a loop of thread weighing only the one-sixteenth of a grain (4·05 mg) caused them to become / greatly curved in under 4 hrs. When the loop was removed, the petioles straightened themselves in about the same time. The petiole (*e*) was rather less sensitive; and in another specimen, in which the corresponding petiole bore rather larger leaflets, a loop of thread weighing one-eighth of a grain did not cause curvature until 18 hrs had elapsed. Loops of thread weighing one-fourth of a grain, left

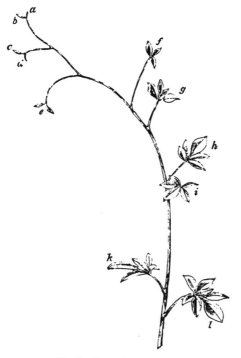

Fig. 8   *Corydalis claviculata*
Leaf-tendril, of natural size

suspended on the lower / petioles (*f* to *l*) during several days, produced no effect. Yet the three petioles *f*, *g*, and *h* were not quite insensible, for when left in contact with a stick for a day or two they slowly curled round it. Thus the sensibility of the petiole gradually diminishes from the tendril-like extremity to the base. The internodes of the stem are not at all sensitive, which makes Mohl's statement that they are sometimes converted into tendrils the more surprising, not to say improbable.

The whole leaf, whilst young and sensitive, stands almost vertically upwards, as we have seen to be the case with many tendrils. It is in continual movement, and one that I observed swept at an average rate of about 2 hrs for each revolution, large, though irregular, ellipses, which were sometimes narrow, sometimes broad, with their longer axes directed to different points of the compass. The young inter-nodes, likewise revolved irregularly in ellipses or spires; so that by

79

these combined movements a considerable space was swept for a support. If the terminal and attenuated portion of a petiole fails to seize any object, it ultimately bends downwards and inwards, and soon loses all irritability and power of movement. This bending down differs much in nature from that which occurs with the extremities of the young leaves in many species of clematis; for these, when thus bent downwards or hooked, first acquire their full degree of sensitiveness.

Dicentra thalictrifolia. In this allied plant the / metamorphoses of the terminal leaflets is complete, and they are converted into perfect tendrils. Whilst the plant is young, the tendrils appear like modified branches, and a distinguished botanist thought that they were of this nature; but in a full-grown plant there can be no doubt, as I am assured by Dr Hooker, that they are modified leaves. When of full size, they are above 5 inches in length; they bifurcate twice, thrice, or even four times; their extremities are hooked and blunt. All the branches of the tendrils are sensitive on all sides, but the basal portion of the main stem is only slightly so. The terminal branches when lightly rubbed with a twig became curved in the course of from 30 m to 42 m, and straightened themselves in between 10 hrs and 20 hrs. A loop of thread weighing one-eighth of a grain plainly caused the thinner branches to bend, as did occasionally a loop weighing one-sixteenth of a grain; but this latter weight, though left suspended, was not sufficient to cause a permanent flexure. The whole leaf with its tendril, as well as the young upper internodes, revolves vigorously and quickly, though irregularly, and thus sweeps a wide space. The figure traced on a bell-glass was either an irregular spire or a zigzag line. The nearest approach to an ellipse was an elongated figure of 8, with one end a little open, and this was completed in 1 hr 53 m. During a period of 6 hrs 17 m another shoot made a complex figure, apparently representing three and a half ellipses. When the lower part of the petiole bearing the leaflets / was securely fastened, the tendril itself described similar but much smaller figures.

This species climbs well. The tendrils after clasping a stick become thicker and more rigid; but the blunt hooks do not turn and adapt themselves to the supporting surface, as is done in so perfect a manner by some Bignoniaceae and Cobaea. The tendrils of young plants, two or three feet in height, are only half the length of those borne by the same plant when grown taller, and they do not contract spirally after clasping a support, but only become slightly flexuous. Full-sized tendrils, on the other hand, contract spirally, with the exception of the thick basal portion. Tendrils which have caught nothing simply bend downwards and inwards, like the extremities of the leaves of the *Corydalis claviculata*. But in all cases the petiole after a time is angularly and abruptly bent downwards like that of Eccremocarpus. /

CHAPTER IV

TENDRIL-BEARERS *continued*

Cucurbitaceae – Homologous nature of the tendrils – *Echinocystis lobata*, remarkable movements of the tendrils to avoid seizing the terminal shoot – Tendrils not excited by contact with other tendrils or by drops of water – Undulatory movement of the extremity of the tendril – Hanburya, adherent discs – Vitaceae – Gradation between the flower-peduncles and tendrils of the vine – Tendrils of the Virginian Creeper turn from the light, and after contact, develop adhesive discs – Sapindaceae – Passifloraceae – *Passiflora gracilis* – Rapid revolving movement and sensitiveness of the tendrils – Not sensitive to the contact of other tendrils or of drops of water – Spiral contraction of tendrils – Summary on the nature and action of tendrils

CUCURBITACEAE

The tendrils in this family have been ranked by competent judges as modified leaves, stipules, or branches; or as partly a leaf and partly a branch. De Candolle believes that the tendrils differ in their homological nature in two of the tribes.[1] From facts recently adduced, Mr Berkeley thinks that Payer's view is the most probable, namely, that the tendril is 'a separate portion of the leaf itself'; but much may be said in favour of the belief that it is a modified flower-peduncle.[2] /

Echinocystis lobata. Numerous observations were made on this plant (raised from seed sent me by Professor Asa Gray), for the spontaneous revolving movements of the internodes and tendrils were first observed by me in this case, and greatly perplexed me. My observations may

[1] I am indebted to Professor Oliver for information on this head. In the *Bulletin de la Société Botanique de France*, 1857, there are numerous discussions on the nature of the tendrils in this family.

[2] *Gardeners' Chronicle*, 1864, p. 721. From the affinity of the Cucurbitaceae to the Passifloraceae, it might be argued that the tendrils of the former are modified flower-peduncles, as is certainly the case with those of Passion-flowers. / Mr R. Holland (Hardwicke's *Science-Gossip*, 1865, p. 105) states that 'a cucumber grew, a few years ago in my own garden, where one of the short prickles upon the fruit had grown out into a long, curled tendril'.

now be much condensed. I observed thirty-five revolutions of the internodes and tendrils; the slowest rate was 2 hrs, and the average rate, with no great fluctuations, 1 hr 40 m. Sometimes I tied the internodes, so that the tendrils alone moved; at other times I cut off the tendrils whilst very young, so that the internodes revolved by themselves; but the rate was not thus affected. The course generally pursued was with the sun, but often in an opposite direction. Sometimes the movement during a short time would either stop or be reversed; and this apparently was due to interference from the light, as, for instance, when I placed a plant close to a window. In one instance, an old tendril, which had nearly ceased revolving, moved in one direction, whilst a young tendril above moved in an opposite course. The two uppermost internodes alone revolve; and as soon as the lower one grows old, only its upper part continues to move. The ellipses or circles swept by the summits of the internodes are about three inches in diameter; whilst those swept by the tips of the / tendrils, are from 15 to 16 inches in diameter. During the revolving movement, the internodes become successively curved to all points of the compass; in one part of their course they are often inclined, together with the tendrils, at about 45° to the horizon, and in another part stand vertically up. There was something in the appearance of the revolving internodes which continually gave the false impression that their movement was due to the weight of the long and spontaneously revolving tendril; but, on cutting off the latter with sharp scissors, the top of the shoot rose only a little, and went on revolving. This false appearance is apparently due to the internodes and tendrils all curving and moving harmoniously together.

A revolving tendril, though inclined during the greater part of its course at an angle of about 45° (in one case of only 37°) above the horizon, stiffened and straightened itself from tip to base in a certain part of its course, thus becoming nearly or quite vertical. I witnessed this repeatedly; and it occurred both when the supporting internodes were free and when they were tied up; but was perhaps most conspicuous in the latter case, or when the whole shoot happened to be much inclined. The tendril forms a very acute angle with the projecting extremity of the stem or shoot; and the stiffening always occurred as the tendril approached, and had to pass over the shoot in its circular course. If it had not possessed and exercised this curious power, it would infallibly have struck against the extremity of the shoot and been / arrested. As soon as the tendril with its three branches begins to stiffen itself in this manner and to rise from an inclined into a vertical position, the revolving motion becomes more rapid; and as soon as the tendril has succeeded in passing over the extremity of the shoot or point of difficulty, its motion, coinciding with that from its weight, often causes it to fall into its previously inclined position so quickly, that the apex could be seen travelling like the minute hand of a gigantic clock.

The tendrils are thin, from 7 to 9 inches in length, with a pair of short lateral branches rising not far from the base. The tip is slightly and permanently curved, so as to act to a limited extent as a hook. The concave side of the tip is highly sensitive to a touch; but not so the convex side, as was likewise observed to be the case with other species of the family by Mohl (p. 65). I repeatedly proved this difference by lightly rubbing four or five times the convex side of one tendril, and only once or twice the concave side of another tendril, and the latter alone, curled inwards. In a few hours afterwards, when the tendrils which had been rubbed on the concave side had straightened themselves, I reversed the process of rubbing, and always with the same result. After touching the concave side, the tip becomes sensibly curved in one or two minutes; and subsequently, if the touch has been at all rough, it coils itself into a helix. But the helix will, after a time, straighten itself, and be again ready to act. A loop of thin thread only one-sixteenth of / a grain in weight caused a temporary flexure. The lower part was repeatedly rubbed rather roughly, but no curvature ensued; yet this part is sensitive to prolonged pressure, for when it came into contact with a stick, it would slowly wind round it.

One of my plants bore two shoots near together, and the tendrils were repeatedly drawn across one another, but it is a singular fact that they did not once catch each other. It would appear as if they had become habituated to contact of this kind, for the pressure thus caused must have been much greater than that caused by a loop of soft thread weighing only the one-sixteenth of a grain. I have, however, seen several tendrils of *Bryonia dioica* interlocked, but they subsequently released one another. The tendrils of the Echinocystis are also habituated to drops of water or to rain; for artificial rain made by violently flirting a wet brush over them produced not the least effect.

The revolving movement of a tendril is not stopped by the curving of its extremity after it has been touched. When one of the lateral branches has firmly clasped an object, the middle branch continues to revolve. When a stem is bent down and secured, so that the tendril depends but is left free to move, its previous revolving movement is nearly or quite stopped; but it soon begins to bend upwards, and as soon as it has become horizontal the revolving movement recommences. I tried this four times; the tendril generally rose to a horizontal position in an hour or an hour and / a half; but in one case, in which a tendril depended at an angle of 45° beneath the horizon, the uprising took two hours; in half an hour afterwards it rose to 23° above the horizon and then recommenced revolving. This upward movement is independent of the action of light, for it occurred twice in the dark, and on another occasion the light came in on one side alone. The movement no doubt is guided by opposition to the force of gravity, as in the case of the ascent of the plumules of germinating seeds.

A tendril does not long retain its revolving power; and as soon as this is lost, it bends downwards and contracts spirally. After the revolving movement has ceased, the tip still retains for a short time its sensitiveness to contact, but this can be of little or no use to the plant.

Though the tendril is highly flexible, and though the extremity travels, under favourable circumstances, at about the rate of an inch in two minutes and a quarter, yet its sensitiveness to contact is so great that it hardly ever fails to seize a thin stick placed in its path. The following case surprised me much: I placed a thin, smooth, cylindrical stick (and I repeated the experiment seven times) so far from a tendril, that its extremity could only curl half or three-quarters round the stick; but I always found that the tip managed in the course of a few hours to curl twice or even thrice round the stick. I at first thought that this was due to rapid growth on the outside; but by coloured points and measurements I proved that / there had been no sensible increase of length within the time. When a stick, flat on one side, was similarly placed, the tip of the tendril could not curl beyond the flat surface, but coiled itself into a helix, which, turning to one side, lay flat on the little flat surface of wood. In one instance a portion of tendril three-quarters of an inch in length was thus dragged on to the flat surface by the coiling in of the helix. But the tendril thus acquires a very insecure hold, and generally after a time slips off. In one case alone the helix subsequently uncoiled itself, and the tip then passed round and clasped the stick. The formation of the helix on the flat side of the stick apparently shows us that the continued striving of the tip to curl itself closely inwards gives the force which drags the tendril round a smooth cylindrical stick. In this latter case, whilst the tendril was slowly and

85

quite insensibly crawling onwards, I observed several times through a
lens that the whole surface was not in close contact with the stick; and I
can understand the onward progress only by supposing that the
movement is slightly undulatory or vermicular, and that the tip
alternately straightens itself a little and then again curls inwards. It
thus drags itself onwards by an insensibly slow, alternate movement,
which may be compared to that of a strong man suspended by the ends
of his fingers to a horizontal pole, who works his fingers onwards until
he can grasp the pole with the palm of his hand. However this may be,
the fact is certain that a tendril which has caught a round stick / with its
extreme point, can work itself onwards until it has passed twice or even
thrice round the stick, and has permanently grasped it.

Hanburya Mexicana. The young internodes and tendrils of this
anomalous member of the family, revolve in the same manner and at
about the same rate as those of the *Echinocystis*. The stem does not
twine, but can ascend an upright stick by the aid of its tendrils. The
concave tip of the tendril is very sensitive; after it had become rapidly
coiled into a ring owing to a single touch, it straightened itself in 50 m.
The tendril, when in full action, stands vertically up, with the
projecting extremity of the young stem thrown a little on one side, so
as to be out of the way; but the tendril bears on the inner side, near its
base, a short rigid branch, which projects out at right angles like a
spur, with the terminal half bowed a little downwards. Hence, as the
main vertical branch revolves, the spur, from its position and rigidity,
cannot pass over the extremity of the shoot, in the same curious
manner as do the three branches of the tendril of the *Echinocystis*,
namely, by stiffening themselves at the proper point. The spur is
therefore pressed laterally against the young stem in one part of the
revolving course, and thus the sweep of the lower part of the main
branch is much restricted. A nice case of co-adaptation here comes into
play: in all the other tendrils observed by me, the several branches
become sensitive at the same period: had this been the case with the
Hanburya, the inwardly directed, spur-like branch, from being /
pressed, during the revolving movement, against the projecting end of
the shoot, would infallibly have seized it in a useless or injurious
manner. But the main branch of the tendril, after revolving for a time
in a vertical position, spontaneously bends downwards; and in doing
so, raises the spur-like branch, which itself also curves upwards; so that
by these combined movements it rises above the projecting end of the

shoot, and can now move freely without touching the shoot; and now it first becomes sensitive.

The tips of both branches, when they come into contact with a stick, grasp it like any ordinary tendril. But in the course of a few days, the lower surface swells and becomes developed into a cellular layer, which adapts itself closely to the wood, and firmly adheres to it. This layer is analogous to the adhesive discs formed by the extremities of the tendrils of some species of *Bignonia* and of *Ampelopsis*; but in the *Hanburya* the layer is developed along the terminal inner surface, sometimes for a length of 1¾ inches, and not at the extreme tip. The layer is white, whilst the tendril is green, and near the tip it is sometimes thicker than the tendril itself; it generally spreads a little beyond the sides of the tendril, and is fringed with free elongated cells, which have enlarged globular or retort-shaped heads. This cellular layer apparently secretes some resinous cement; for its adhesion to the wood was not lessened by an immersion of 24 hrs in alcohol or water, but was quite loosened by a similar immersion in ether or turpentine. / After a tendril has once firmly coiled itself round a stick, it is difficult to imagine of what use the adhesive cellular layer can be. Owing to the spiral contraction which soon ensues, the tendrils were never able to remain, excepting in one instance, in contact with a thick post or a nearly flat surface; if they had quickly become attached by means of the adhesive layer, this would evidently have been of service to the plant.

The tendrils of *Bryonia dioica, Cucurbita ovifera*, and *Cucumis sativa* are sensitive and revolve. Whether the internodes likewise revolve I did not observe. In *Anguria Warscewiczii*, the internodes, though thick and stiff, revolve: in this plant the lower surfaces of the tendril, some time after clasping a stick, produces a coarsely cellular layer or cushion, which adapts itself closely to the wood, like that formed by the tendril of the Hanburya; but it is not in the least adhesive. In *Zanonia Indica*, which belongs to a different tribe of the family, the forked tendrils and the internodes revolve in periods between 2 hrs 8 m and 3 hrs 35 m, moving against the sun.

VITACEAE

In this family and in the two following, namely, the Sapindaceae and Passifloraceae, the tendrils are modified flower-peduncles; and are

87

therefore axial in their nature. In this respect they differ from all those previously described, with the exception, perhaps, of the Cucurbitacea. The homological nature, however, of a tendril seems to make no difference in its action. /

Vitis vinifera. The tendril is thick and of great length; one from a vine growing out of doors and not vigorously, was 16 inches long. It consists of a peduncle (A), bearing two branches which diverge equally from it. One of the branches (B) has a scale at its base; it is always, as far as I have seen, longer than the other and often bifurcates. The

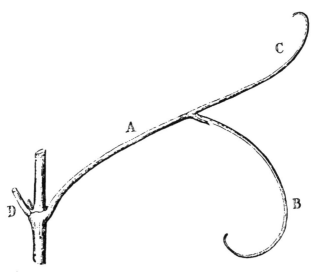

Fig. 9   Tendril of the vine
A. Peduncle of tendril, B. Longer branch, with a scale at its base, C. Shorter branch, D. Petiole of the opposite leaf

branches when rubbed become curved, and subsequently straighten themselves. After a tendril has clasped any object with its extremity, it contracts spirally; but this does not occur (Palm, p. 56) when no object has been seized. The tendrils move spontaneously / from side to side; and on a very hot day, one made two elliptical revolutions, at an average rate of 2 hrs 15 m. During these movements a coloured line, painted along the convex surface, appeared after a time on one side, then on the concave side, then on the opposite side, and lastly again on the convex side. The two branches of the same tendril have independent movements. After a tendril has spontaneously revolved for a time, it bends from the light towards the dark: I do not state this on my own authority, but on that of Mohl and Dutrochet. Mohl (p. 77) says that in a vine planted against a wall the tendrils point towards it, and in a vineyard generally more or less to the north.

The young internodes revolve spontaneously; but the movement is unusually slight. A shoot faced a window, and I traced its course on the glass during two perfectly calm and hot days. On one of these days it described, in the course of ten hours, a spire, representing two and a half ellipses. I also placed a bell-glass over a young Muscat grape in the hothouse, and it made each day three or four very small oval revolutions; the shoot moving less than half an inch from side to side. Had it not made at least three revolutions whilst the sky was uniformly overcast, I should have attributed this slight degree of movement to the varying action of the light. The extremity of the stem is more or less bent downwards, but it never reverses its curvature, as so generally occurs with twining plants. /

Various authors (Palm, p. 55; Mohl, p. 45; Lindley, etc.) believe that the tendrils of the vine are modified flower-peduncles. I here give a drawing (fig. 10) of the ordinary state of a young flower-stalk: it consists of the 'common peduncle' (A); of the 'flower-tendril' (B), which is represented as having caught a twig; and of the 'sub-peduncle' (C) bearing the flower-buds. The whole moves spontaneously, like a true tendril, but in a less degree; the movement, / however, is greater when the sub-peduncle (C) does not bear many flower-buds. The common peduncle (A) has not the power of clasping a support, nor has the corresponding part of a true tendril. The flower-tendril (B) is always longer than the sub-peduncle (C) and has a scale at its base; it sometimes bifurcates, and therefore corresponds in every detail with the longer scale-bearing branch (B, fig. 9) of the true tendril. It is, however, inclined backwards from the sub-peduncle (C), or stands at right angles with it, and is thus adapted to aid in carrying the future bunch of grapes. When rubbed, it curves and subsequently straightens itself; and it can, as is shown in the drawing, securely clasp a support. I have seen an object as soft as a young vine-leaf caught by one.

The lower and naked part of the sub-peduncle (C) is likewise slightly sensitive to a rub, and I have seen it bent round a stick and even partly round a leaf with which it had come into contact. That the sub-peduncle has the same nature as the corresponding branch of an ordinary tendril, is well shown when it bears only a few flowers; for in

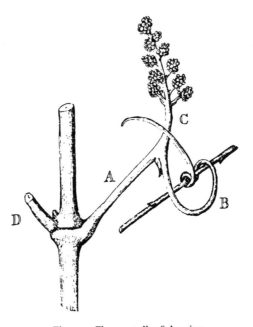

Fig. 10   Flower-stalk of the vine
A. Common peduncle, B. Flower-tendril, with a scale at its base, C. Sub-peduncle,
bearing the flower-buds, D. Petiole of the opposite leaf

this case it becomes less branched, increases in length, and gains both
in sensitiveness and in the power of spontaneous movement. I have
twice seen sub-peduncles which bore from thirty to forty flower-buds,
and which had become considerably elongated and were completely
wound round sticks, exactly like true tendrils. The whole length of
another sub-peduncle, bearing only / eleven flower-buds, quickly
became curved when slightly rubbed; but even this scanty number of
flowers rendered the stalk less sensitive than the other branch, that is,
the flower-tendril; for the latter after a lighter rub became curved
more quickly and in a greater degree. I have seen a sub-peduncle
thickly covered with flower-buds, with one of its higher lateral
branchlets bearing from some cause only two buds; and this one
branchlet had become much elongated and had spontaneously caught
hold of an adjoining twig; in fact, it formed a little sub-tendril. The
increasing length of the sub-peduncle (C) with the decreasing number
of the flower-buds is a good instance of the law of compensation. In
accordance with this same principle, the true tendril as a whole is

always longer than the flower-stalk; for instance, on the same plant, the longest flower-stalk (measured from the base of the common peduncle to the tip of the flower-tendril) was 8½ inches in length, whilst the longest tendril was nearly double this length, namely 16 inches.

The gradations from the ordinary state of a flower-stalk, as represented in the drawing (fig. 10), to that of a true tendril (fig. 9) are complete. We have seen that the sub-peduncle (C), whilst still bearing from thirty to forty flower-buds, sometimes becomes a little elongated and partially assumes all the characters of the corresponding branch of a true tendril. From this state we can trace every stage till we come to a full-sized perfect tendril, bearing on the branch / which corresponds with the sub-peduncle one single flower-bud! Hence there can be no doubt that the tendril is a modified flower-peduncle.

Another kind of gradation well deserves notice. Flower-tendrils (B, fig. 10) sometimes produce a few flower-buds. For instance, on a vine growing against my house, there were thirteen and twenty-two flower-buds respectively on two flower-tendrils, which still retained their characteristic qualities of sensitiveness and spontaneous movement, but in a somewhat lessened degree. On vines in hothouses, so many flowers are occasionally produced on the flower-tendrils that a double bunch of grapes is the result; and this is technically called by gardeners a 'cluster'. In this state the whole bunch of flowers presents scarcely any resemblance to a tendril; and, judging from the facts already given, it would probably possess little power of clasping a support, or of spontaneous movement. Such flower-stalks closely resemble in structure those borne by Cissus. This genus, belonging to the same family of the Vitaceae, produces well-developed tendrils and ordinary bunches of flowers; but there are no gradations between the two states. If the genus Vitis had been unknown, the boldest believer in the modification of species would never have surmised that the same individual plant, at the same period of growth, would have yielded every possible gradation between ordinary flower-stalks for the support of the flowers and fruit, and tendrils used exclusively for climbing. But the vine clearly gives us such a case; and it / seems to me as striking and curious an instance of transition as can well be conceived.

Cissus discolor. The young shoots show no more movement than can be accounted for by daily variations in the action of the light. The tendrils, however, revolve with much regularity, following the sun; and, in the plants observed by me, swept circles of about 5 inches in diameter. Five circles were completed in the following times: 4 hrs 45 m, 4 hrs 50 m, 4 hrs 45 m, 4 hrs 30 m, and 5 hrs. The same tendril continues to revolve during three or four days. The tendrils are from 3½ to 5 inches in length. They are formed of a long footstalk, bearing two short branches, which in old plants again bifurcate. The two branches are not of quite equal length; and as with the vine, the longer one has a scale at its base. The tendril stands vertically upwards; the extremity of the shoot being bent abruptly downwards, and this position is probably of service to the plant by allowing the tendril to revolve freely and vertically.

Both branches of the tendril, whilst young, are highly sensitive. A touch with a pencil, so gentle as only just to move a tendril borne at the end of a long flexible shoot, sufficed to cause it to become perceptibly curved in four or five minutes. It became straight again in rather above one hour. A loop of soft thread weighing one-seventh of a grain (9·25 mg) was thrice tried, and each time caused the tendril to become curved in 30 or 40 m. Half this weight produced no effect. The long footstalk is much less / sensitive, for a slight rubbing produced no effect, although prolonged contact with a stick caused it to bend. The two branches are sensitive on all sides, so that they converge if touched on their inner sides, and diverge if touched on their outer sides. If a branch be touched at the same time with equal force on opposite sides, both sides are equally stimulated and there is no movement. Before examining this plant, I had observed only tendrils which are sensitive on one side alone, and these when lightly pressed between the finger and thumb become curved; but on thus pinching many times the tendrils of the Cissus no curvature ensued, and I falsely inferred at first that they were not at all sensitive.

Cissus antarcticus. The tendrils on a young plant were thick and straight, with the tips a little curved. When their concave surfaces were rubbed, and it was necessary to do this with some force, they very slowly became curved, and subsequently straight again. They are therefore much less sensitive than those of the last species; but they made two revolutions, following the sun, rather more rapidly, viz., in 3 hrs 30 m and 4 hrs. The internodes do not revolve.

Ampelopsis hederacea (Virginian Creeper). The internodes apparently do not move more than can be accounted for by the varying action of the light. The tendrils are from 4 to 5 inches in length, with the main stem sending off several lateral branches, which have their tips curved, as may be seen in the upper figure (fig. 11). They exhibit no true spontaneous revolving/movement, but turn, as was long ago observed by Andrew Knight,[3] from the light to the dark. I have seen several tendrils move in less than 24 hours, through an angle of 180° to the dark side of a case in which a plant was placed, but the movement is sometimes much slower. The several lateral branches often move independently of one another, and sometimes irregularly, without any apparent cause. These tendrils are less sensitive to a touch than any others observed by me. By gentle but repeated rubbing with a twig, the lateral branches, but not the main stem, became in the course of three or four hours slightly curved; but they seemed to have hardly any power of again straightening themselves. The tendrils of a plant which had crawled over a large box-tree clasped several of the branches; but I have repeatedly seen that they will withdraw themselves after seizing a stick. When they meet with a flat surface of wood or a wall (and this is evidently what they are adapted for), they turn all their branches towards it, and, spreading them widely apart, bring their hooked tips laterally into contact with it. In effecting this, the several branches, after touching the surface, often rise up, place themselves in a new position, and again come down into contact with it.

In the course of about two days after a tendril has arranged its branches so as to press on any surface, the curved tips swell, become bright red, and form on / their undersides the well-known little discs or cushions with which they adhere firmly. In one case the tips were slightly swollen in 38 hrs after coming into contact with a brick; in another case they were considerably swollen in 48 hrs, and in an additional 24 hrs were firmly attached to a smooth board; and lastly, the tips of a younger tendril not only swelled but became attached to a stuccoed wall in 42 hrs. These adhesive discs resemble, except in colour and in being larger, those of *Bignonia capreolata*. When they were developed in contact with a ball of tow, the fibres were separately enveloped, but not in so effective a manner as by *B. capreolata*. Discs are never developed, as far as I have seen, without the stimulus of at least temporary contact with some object.[4] They are

[3] *Trans. Phil. Soc.*, 1812, p. 314.

[4] Dr M'Nab remarks (*Trans. Bot. Soc. Edinburgh*, vol. xi, p. 292) that the tendrils of *Amp. Veitchii* bear small globular discs before they have come into contact with any object; and I have since observed the same fact. These discs, however, increase greatly in size, if they press against and adhere to any surface. The tendrils, therefore, of one species of Ampelopsis require the stimulus of contact for the first development of their discs, whilst those of another species do not need any such stimulus. We have seen an exactly parallel case with two species of Bignoniaceae.

generally first formed on one side of the curved tip, the whole of which often becomes so much changed in appearance, that a line of the original green tissue can be traced only along the concave surface. When, however, a tendril has clasped a cylindrical stick, an irregular rim or disc is sometimes formed along the inner surface at some little distance from the curved / tip; this was also observed (p. 71) by Mohl. The discs consist of enlarged cells, with smooth projecting hemispherical surfaces, coloured red; they are at first gorged with fluid (see section given by Mohl, p. 70), but ultimately become woody.

As the discs soon adhere firmly to such smooth surfaces as planed or painted wood, or to the polished leaf of the ivy, this alone renders it probable that some cement is secreted, as has been asserted to be the case (quoted by Mohl, p. 71) by Malpighi. I removed a number of discs formed during the previous year from a stuccoed wall, and left them during many hours, in warm water, diluted acetic acid and alcohol; but the attached grains of silex were not loosened. Immersion in sulphuric ether for 24 hrs loosened them much, but warmed essential oils (I tried oil of thyme and peppermint) completely released every particle of stone in the course of a few hours. This seems to prove that some resinous cement is secreted. The quantity, however, must be small; for when a plant ascended a thinly whitewashed wall, the discs adhered firmly to the whitewash; but as the cement never penetrated the thin layer, they were easily withdrawn, together with little scales of the whitewash. It must not be supposed that the attachment is effected exclusively by the cement; for the cellular outgrowth completely envelopes every minute and irregular projection, and insinuates itself into every crevice.

A tendril which has not become attached to any body, does not contract spirally; and in course of a / week or two shrinks into the finest thread, withers and drops off. An attached tendril, on the other hand, contracts spirally, and thus becomes highly elastic, so / that when the main footstalk is pulled the strain is distributed equally between all the attached discs. For a few days after the attachment of the discs, the tendril remains weak and brittle, but it rapidly increases in thickness and acquires great strength. During the following winter it ceases to live, but adheres firmly in a dead state both to its own stem and to the surface of attachment. In the accompanying diagram (fig. 11) we see the difference between a tendril (B) some weeks after its attachment to a wall, with one (A) from the same plant fully grown but unattached. That the change in the nature of the tissues, as well as the spiral

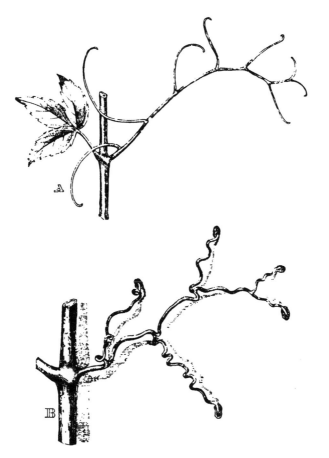

Fig. 11   *Ampelopsis hederacea*
A. Tendril fully developed, with a young leaf on the opposite side of the stem, B. Older tendril, several weeks after its attachment to a wall, with the branches thickened and spirally contracted, and with the extremities developed into discs. The unattached branches of this tendril have withered and dropped off.

contraction, are consequent on the formation of the discs, is well shown by any lateral branches which have not become attached; for these in a week or two wither and drop off, in the same manner as does the whole tendril if unattached. The gain in strength and durability in a tendril after its attachment is something wonderful. There are tendrils now adhering to my house which are still strong, and have been exposed to the weather in a dead state for fourteen or fifteen years. One single lateral branchlet of a tendril, estimated to be at least ten years old, was still elastic and supported a weight of exactly two pounds. The whole tendril had five disc-bearing branches of equal thickness and apparently of equal strength; so that after having been exposed during ten years to the weather, it would probably have resisted a strain of ten pounds!

SAPINDACEAE

Cardiospermum halicacabum. In this / family, as in the last, the tendrils are modified flower-peduncles. In the present plant the two lateral branches of the main flower-peduncle have been converted into a pair of tendrils, corresponding with the single 'flower-tendril' of the common vine. The main peduncle is thin, stiff, and from 3 to 4½ inches in length. Near the summit, above two little bracts, it divides

Fig. 12   *Cardiospermum halicacabum*
Upper part of the flower-peduncle with its two tendrils

into three branches. The middle one divides and redivides, and bears the flowers; ultimately it grows half as long again as the two other modified branches. These latter are the tendrils; they are at first thicker and longer than the middle branch, but never become more than an inch in length. They taper to a point and are flattened, with the lower clasping surface destitute of hairs. At first they project straight up; but soon diverging, spontaneously curl downwards so as to become symmetrically and elegantly hooked, as represented in the diagram. They are now, whilst the flower-buds are still small, ready for action. /

The two or three upper internodes, whilst young, steadily revolve; those on one plant made two circles, against the course of the sun, in

3 hrs 12 m; in a second plant the same course was followed, and the two circles were completed in 3 hrs 41 m; in a third plant, the internodes followed the sun and made two circles in 3 hrs 47 m. The average rate of these six revolutions was 1 hr 46 m. The stem shows no tendency to twine spirally round a support; but the allied tendril-bearing genus Paullinia is said (Mohl, p. 4) to be a twiner. The flower-peduncles, which stand up above the end of the shoot, are carried round and round by the revolving movement of the internodes; and when the stem is securely tied, the long and thin flower-peduncles themselves are seen to be in continued and sometimes rapid move-ment from side to side. They sweep a wide space, but only occasionally revolve in a regular elliptical course. By the combined movements of the internodes and peduncles, one of the two short hooked tendrils, sooner or later, catches hold of some twig or branch, and then it curls round and securely grasps it. These tendrils are, however, but slightly sensitive; for by rubbing their under surface only a slight movement is slowly produced. I hooked a tendril on to a twig; and in 1 hr 45 m it was curved considerably inwards; in 2 hrs 30 m it formed a ring; and in from 5 to 6 hours from being first hooked, it closely grasped the stick. A second tendril acted at nearly the same rate; but I observed one that took 24 hours before it curled twice round a / thin twig. Tendrils which have caught nothing, spontaneously curl up to a close helix after the interval of several days. Those which have curled round some object, soon become a little thicker and tougher. The long and thin main peduncle, though spontaneously moving, is not sensitive and never clasps a support. Nor does it ever contract spirally,[5] although a contraction of this kind apparently would have been of service to the plant in climbing. Nevertheless it climbs pretty well without this aid. The seed-capsules though light, are of enormous size (hence its English name of balloon-vine), and as two or three are carried on the same peduncle, the tendrils rising close to them may be of service in preventing their being dashed to pieces by the wind. In the hothouse the tendrils served simply for climbing.

The position of the tendrils alone suffices to show their homological nature. In two instances one of two tendrils produced a flower at its tip; this, however, did not prevent its acting properly and curling

[5] Fritz Müller remarks (ibid., p. 348) that a related genus, Serjania, differs from Cardiospermum in bearing only a single tendril; and that the common peduncle contracts spirally, when, as frequently happens, the tendril has clasped the plant's own stem.

round a twig. In a third case both lateral branches which ought to have been modified into tendrils, produced flowers like the central branch, and had quite lost their tendril-structure.

I have seen, but was not enabled carefully to observe, only one other climbing Sapindaceous plant, namely, / Paullinia. It was not in flower, yet bore long forked tendrils. So that, Paullinia, with respect to its tendrils, appears to bear the same relation to Cardiospermum that Cissus does to Vitis.

PASSIFLORACEAE

After reading the discussion and facts given by Mohl (p. 47) on the nature of the tendrils in this family, no one can doubt that they are modified flower-peduncles. The tendrils and the flower-peduncles rise close side by side; and my son, William E. Darwin, made sketches for me of their earliest state of development in the hybrid *P. floribunda*. The two organs appear at first as a single papilla, which gradually divides; so that the tendril appears to be a modified branch of the flower-peduncle. My son found one very young tendril surmounted by traces of floral organs, exactly like those on the summit of the true flower-peduncle at the same early age.

Passiflora gracilis. This well-named, elegant, annual species differs from the other members of the group observed by me, in the young internodes having the power of revolving. It exceeds all the other climbing plants which I have examined, in the rapidity of its movements, and all tendril-bearers in the sensitiveness of the tendrils. The internode which carries the upper active tendril and which likewise carries one or two younger immature internodes, made three revolutions, following the sun, at an average rate of 1 hr 4 m; it then made, the day becoming very hot, three other revolutions at an average rate of between 57 and 58 m; so that the average of all six revolutions was / 1 hr 1 m. The apex of the tendril describes elongated ellipses, sometimes narrow and sometimes broad, with their longer axes inclined in slightly different directions. The plant can ascend a thin upright stick by the aid of its tendrils; but the stem is too stiff for it to twine spirally round it, even when not interfered with by the tendrils, these having been successively pinched off at an early age.

When the stem is secured, the tendrils are seen to revolve in nearly the same manner and at the same rate as the internodes.[6] The tendrils are very thin, delicate, and straight, with the exception of the tips, which are a little curved; they are from 7 to 9 inches in length. A half-grown tendril is not sensitive; but when nearly full-grown they are extremely sensitive. A single delicate touch on the concave surface of the tip soon caused one to curve; and in 2 minutes it formed an open helix. A loop of soft thread weighing ⅓₂nd of a grain (2·02 mg) placed most gently on the tip, thrice caused distinct curvature. A bent bit of thin platina wire weighing only ¹⁄₅₀th of a grain (1·23 mg) twice produced the same effect; but this latter weight, when left suspended, did not suffice to cause a permanent curvature. These trials were made under a bell-glass, so that the loops of thread and wire were / not agitated by the wind. The movement after a touch is very rapid: I took hold of the lower part of several tendrils, and then touched their concave tips with a thin twig and watched them carefully through a lens; the tips evidently began to bend after the following intervals – 31, 25, 32, 31, 28, 39, 31, and 30 seconds; so that the movement was generally perceptible in half a minute after a touch; but on one occasion it was distinctly visible in 25 seconds. One of the tendrils which thus became bent in 31 seconds, had been touched two hours previously and had coiled into a helix; so that in this interval it had straightened itself and had perfectly recovered its irritability.

To ascertain how often the same tendril would become curved when touched, I kept a plant in my study, which from being cooler than the hothouse was not very favourable for the experiment. The extremity was gently rubbed four or five times with a thin stick, and this was done as often as it was observed to have become nearly straight again after having been in action; and in the course of 54 hrs it answered to the stimulus 21 times, becoming each time hooked or spiral. On the last occasion, however, the movement was very slight, and soon afterwards permanent spiral contraction commenced. No trials were made during the night, so that the tendril would perhaps have answered a greater number of times to the stimulus; though, on the other hand, from having no rest it might have become exhausted from so many quickly repeated efforts. /

I repeated the experiment made on the Echinocystis, and placed

[6] Professor Asa Gray informs me that the tendrils of *P. acerifolia* revolve even at a quicker rate than those of *P. gracilis*; four revolutions were completed (the temperature varying from 88°–92° Fahr.) in the following times: 40 m, 45 m, 38½ m, and 46 m. One half-revolution was performed in 15 m.

99

several plants of this Passiflora so close together, that their tendrils were repeatedly dragged over each other; but no curvature ensued. I likewise repeatedly flirted small drops of water from a brush on many tendrils, and syringed others so violently that the whole tendril was dashed about, but they never became curved. The impact from the drops of water was felt far more distinctly on my hand than that from the loops of thread (weighing ⅟₃₂nd of a grain) when allowed to fall on it from a height, and these loops, which caused the tendrils to become curved, had been placed most gently on them. Hence it is clear, that the tendrils either have become habituated to the touch of other tendrils and drops of rain, or that they were from the first rendered sensitive only to prolonged though excessively slight pressure of solid objects, with the exclusion of that from other tendrils. To show the difference in the kind of sensitiveness in different plants and likewise to show the force of the syringe used, I may add that the lightest jet from it instantly caused the leaves of a Mimosa to close; whereas the loop of thread weighing ⅟₃₂nd of a grain, when rolled into a ball and placed gently on the glands at the base of the leaflets of the Mimosa, caused no action.

Passiflora punctata. The internodes do not move, but the tendrils revolve regularly. A half-grown and very sensitive tendril made three revolutions, opposed to the course of the sun, in 3 hrs 5 m, 2 hrs 40 m, and 2 hrs 50 m; perhaps it might have travelled more / quickly when nearly full-grown. A plant was placed in front of a window, and, as with twining stems, the light accelerated the movement of the tendril in one direction and retarded it in the other; the semicircle towards the light being performed in one instance in 15 m less time and in a second instance in 20 m less time than that required by the semicircle towards the dark end of the room. Considering the extreme tenuity of these tendrils, the action of the light on them is remarkable. The tendrils are long, and, as just stated, very thin, with the tip slightly curved or hooked. The concave side is extremely sensitive to a touch – even a single touch causing it to curl inwards; it subsequently straightened itself, and was again ready to act. A loop of soft thread weighing ⅟₁₄th of a grain (4·625 mg) caused the extreme tip to bend; another time I tried to hang the same little loop on an inclined tendril, but three times it slid off; yet this extraordinarily slight degree of friction sufficed to make the tip curl. The tendril, though so sensitive, does not move very quickly after a touch, no conspicuous movement being observable until 5 or 10 m had elapsed. The convex side of the tip is not sensitive to a touch or to a suspended loop of thread. On one occasion I observed a tendril revolving with the convex side of the tip forwards, and in consequence it was not able to clasp a stick, against which it scraped; whereas tendrils revolving with the concave side forward, promptly seize any object in their path.

Passiflora quadrangularis. This is a very distinct / species. The tendrils are thick, long, and stiff; they are sensitive to a touch only on the concave surface towards the extremity. When a stick was placed so that the middle of the tendril came into contact with it, no curvature ensued. In the hothouse a tendril made two revolutions, each in 2 hrs 22 m; in a cool room one was completed in 3 hrs, and a second in 4 hrs. The internodes do not revolve; nor do those of the hybrid *P. floridunda*.

Tacsonia manicata. Here again the internodes do not revolve. The tendrils are moderately thin and long; one made a narrow ellipse in 5 hrs 20 m, and the next day a broad ellipse in 5 hrs 7 m. The extremity being lightly rubbed on the concave surface, became just perceptibly curved in 7 m, distinctly in 10 m, and hooked in 20 m.

We have seen that the tendrils in the last three families, namely, the Vitaceae, Sapindaceae, and Passifloraceae, are modified flower-peduncles. This is likewise the case, according to De Candolle (as quoted by Mohl), with the tendrils of Brunnichia, one of the Polygonaceae. In two or three species of Modecca, one of the Papayaceae, the tendrils, as I hear from Professor Oliver, occasionally bear flower and fruit; so that they are axial in their nature.

The spiral contraction of tendrils

This movement, which shortens the tendrils and renders them elastic, commences in half a day, or in a day or two after their extremities have caught some / object. There is no such movement in any leaf-climber, with the exception of an occasional trace of it in the petioles of *Tropaeolum tricolorum*. On the other hand, the tendrils of all tendril-bearing plants, contract spirally after they have caught an object with the following exceptions. Firstly, *Corydalis claviculata*, but then this plant might be called a leaf-climber. Secondly and thirdly, *Bignonia unguis* with its close allies, and Cardiospermum; but their tendrils are so short that their contraction could hardly occur, and would be quite superfluous. Fourthly, *Smilax aspera* offers a more marked exception, as its tendrils are moderately long. The tendrils of Dicentra, whilst the plant is young, are short and after attachment only become slightly flexuous; in older plants they are longer and then they contract spirally. I have seen no other exceptions to the rule that tendrils, after clasping with their extremities a support, undergo spiral contraction. When, however, the tendril of a plant of which the stem is immovably fixed, catches some fixed object, it does not contract, simply because it cannot; this, however, rarely occurs. In the common pea the lateral branches alone contract, and not the central stem; and with most plants, such as the vine, Passiflora, Bryony, the basal portion never forms a spire.

I have said that in *Corydalis claviculata* the end of the leaf or tendril (for this part may be indifferently so called) does not contract into a spire. The branchlets, however, after they have wound round / thin twigs, become deeply sinuous or zigzag. Moreover the whole end of the petiole or tendril, if it seizes nothing, bends after a time abruptly downwards and inwards showing that its outer surface has gone on growing after the inner surface has ceased to grow. That growth is the chief cause of the spiral contraction of tendrils may be safely admitted, as shown by the recent researches of H. de Vries. I will, however, add one little fact in support of this conclusion.

If the short, nearly straight portion of an attached tendril of *Passiflora gracilis* (and, as I believe, of other tendrils), between the opposed spires, be examined, it will be found to be transversely wrinkled in a conspicuous manner on the outside; and this would naturally follow if the outer side had grown more than the inner side, this part being at the same time forcibly prevented from becoming curved. So again the whole outer surface of a spirally wound tendril becomes wrinkled if it be pulled straight. Nevertheless, as the contraction travels from the extremity of a tendril, after it has been stimulated by contact with a support, down to the base, I cannot avoid doubting, from reasons presently to be given, whether the whole effect ought to be attributed to growth. An unattached tendril rolls itself up into a flat helix, as in the case of Cardiospermum, if the contraction commences at the extremity and is quite regular; but if the continued growth of the outer surface is a little lateral, or if the process begins near the base, the terminal portion cannot be rolled up within the basal portion, and the / tendril then forms a more or less open spire. A similar result follows if the extremity has caught some object, and is thus held fast.

The tendrils of many kinds of plants, if they catch nothing, contract after an interval of several days or weeks into a spire; but in these cases the movement takes place after the tendril has lost its revolving power and hangs down; it has also then partly or wholly lost its sensibility; so that this movement can be of no use. The spiral contraction of unattached tendrils is a much slower process than that of attached ones. Young tendrils which have caught a support and are spirally contracted, may constantly be seen on the same stem with the much older unattached and uncontracted tendrils. In the Echinocystis I have seen a tendril with the two lateral branches encircling twigs and contracted into beautiful spires, whilst the main branch which had caught nothing remained for many days straight. In this plant I once observed a main branch after it had caught a stick become spirally flexuous in 7 hrs, and spirally contracted in 18 hrs. Generally the tendrils of the Echinocystis begin to contract in from 12 hrs to 24 hrs after catching some object; whilst unattached tendrils do not begin to contract until two or three or even more days after all revolving movement has ceased. A full-grown tendril of *Passiflora quadrangularis* which had caught a stick began in 8 hrs to contract, and in 24 hrs formed several spires; a younger tendril, only two-thirds grown, showed the first trace of contraction in√ two days after clasping a stick, and in two more days formed several spires. It appears, therefore, that the contraction does not begin until the tendril is grown to nearly its full length. Another young tendril of about the same age and length as the last did not catch any object; it acquired its full length in four days; in six additional days it first became flexuous, and in two more days formed one complete spire. This first spire was formed towards the basal end, and the contraction steadily but slowly progressed towards the apex; but the whole was not closely wound up into a spire until 21 days had elapsed from the first observation, that is, until 17 days after the tendril had grown to its full length.

The spiral contraction of tendrils is quite independent of their power of spontaneously revolving, for it occurs in tendrils, such as those of *Lathyrus grandiflorus* and *Ampelopsis hederacea*, which do not revolve. It is not necessarily related to the curling of the tips round a support, as we see with the Ampelopsis and *Bignonia capreolata*, in which the development of adherent discs suffices to cause spiral contraction. Yet in some cases this contraction seems connected with the curling or clasping movement, due to contact with a support; for not only does it soon follow this act, but the contraction generally begins close to the curled extremity, and travels downwards to the base. If, however, a tendril be very slack, the whole length almost simultaneously becomes at first flexuous and then spiral. Again, the tendrils of some few plants / never contract spirally unless they have first seized hold of some object; if they catch nothing they hang down, remaining straight, until they wither and drop off: this is the case with the tendrils of Bignonia, which consist of modified leaves, and with those of three genera of the Vitaceae, which are modified flower-peduncles. But in the great majority of cases, tendrils which have never come in contact with any object, after a time contract spirally. All these facts taken together, show that the act of clasping a support and the spiral contraction of the whole length of the tendril, are phenomena not necessarily connected.

The spiral contraction which ensues after a tendril has caught a support is of high service to the plant; hence its almost universal occurrence with species belonging to widely different orders. When a shoot is inclined and its tendril has caught an object above, the spiral contraction drags up the shoot. When the shoot is upright, the growth of the stem, after the tendrils have seized some object above, would leave it slack, were it not for the spiral contraction which draws up the stem as it increases in length. Thus there is no waste of growth, and the stretched stem ascends by the shortest course. When a terminal branchlet of the tendril of Cobaea catches a stick, we have seen how well the spiral contraction successively brings the other branchlets, one after the other, into contact with the stick, until the whole tendril grasps it in an inextricable knot. When a tendril has caught a yielding object, this is sometimes enveloped and / still further secured by the spiral folds, as I have seen with *Passiflora quadrangularis*; but this action is of little importance.

A far more important service rendered by the spiral contraction of the tendrils is that they are thus made highly elastic. As before remarked under Ampelopsis, the strain is thus distributed equally between the several attached branches; and this renders the whole far stronger than it otherwise would be, as the branches cannot break separately. It is this elasticity which protects both branched and simple tendrils from being torn away from their supports during stormy weather. I have more than once gone on purpose during a gale to watch a Bryony growing in an exposed hedge, with its tendrils attached to the surrounding bushes; and as the thick and thin branches were tossed to and fro by the wind, the tendrils, had they not been excessively elastic, would instantly have been torn off and the plant thrown prostrate. But as it was, the Bryony safely rode out the gale, like a ship with two anchors down, and with a long range of cable ahead to serve as a spring as she surges to the storm.

When an unattached tendril contracts spirally, the spire always runs in the same direction from tip to base. A tendril, on the other hand, which has caught a support by its extremity, although the same side is concave from end to end, invariably becomes twisted in one part in one direction, and in another part in the opposite direction; the oppositely turned spires being separated by a short straight portion. This curious / and symmetrical structure has been noticed by several botanists, but has not been sufficiently explained.[7] It occurs without exception with all tendrils which after catching an object contract spirally, but is of course most conspicuous in the longer tendrils. It never occurs with uncaught tendrils; and when this appears to have occurred, it will be found that the tendril had originally seized some object and had afterwards been torn free. Commonly, all the spires at one end of an attached tendril run in one direction, and all those at the other end in the opposite direction, with a single short straight portion in the middle; but I have seen a tendril with the spires alternately turning five times / in opposite directions, with straight pieces between them;

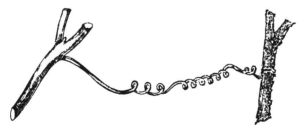

Fig 13   A caught tendril of *Bryonia dioica*, spirally contracted in reversed directions

[7] See M. Isid. Léon in *Bull. Soc. Bot. de France*, vol. v, 1858, p. 680. Dr H. de Vries points out (p. 306) that I have overlooked, in the first edition of this essay, the following sentence by Mohl: 'After a tendril has caught a support, it begins in some days to wind into a spire, which, since the tendril is made fast at both extremities, must of necessity be in some places to the right, in others to the left.' But I am not surprised that this brief sentence, without any further explanation did not attract my attention.

and M. Léon has seen seven or eight such alternations. Whether the spires turn once or more than once in opposite directions, there are as many turns in the one direction as in the other. For instance, I gathered ten attached tendrils of the Bryony, the longest with 33, and the shortest with only 8 spiral turns; and the number of turns in the one direction was in every case the same (within one) as in the opposite direction.

The explanation of this curious little fact is not difficult. I will not attempt any geometrical reasoning, but will give only a practical illustration. In doing this, I shall first have to allude to a point which was almost passed over when treating of twining-plants. If we hold in our left hand a bundle of parallel strings, we can with our right hand turn these round and round, thus imitating the revolving movement of a twining plant, and the strings do not become twisted. But if we hold at the same time a stick in our left hand, in such a position that the string become spirally turned round it, they will inevitaby become twisted. Hence a straight coloured line, painted along the internodes of a twining plant before it has wound round a support, becomes twisted or spiral after it has wound round. I painted a red line on the straight internodes of a Humulus, Mikania, Ceropegia, Convolvulus, and Phaseolus, and saw it become twisted as the plant wound round a stick. It is possible that the stems of some plants by spontaneously turning on / their own axes, at the proper rate and in the proper direction, might avoid becoming twisted; but I have seen no such case.

In the above illustration, the parallel strings were wound round a stick; but this is by no means necessary, for if wound into a hollow coil (as can be done with a narrow slip of elastic paper) there is the same inevitable twisting of the axis. When, therefore, a free tendril coils itself into a spire, it must either become twisted along its whole length (and this never occurs), or the free extremity must turn round as many times as there are spires formed. It was hardly necessary to observe this fact; but I did so by affixing little paper vanes to the extreme points of the tendrils of Echinocystis and *Passiflora quadrangularis*; and as the tendril contracted itself into successive spires, the vane slowly revolved.

We can now understand the meaning of the spires being invariably turned in opposite directions, in tendrils which from having caught some object are fixed at both ends. Let us suppose a caught tendril to make thirty spiral turns all in the same direction; the inevitable result would be that it would become twisted thirty times on its own axis. This twisting would not only require considerable force, but, as I know by trial, would burst the tendril before the thirty turns were completed. Such cases never really occur; for, as already stated, when a tendril has caught a support and is spirally contracted, there are always as many turns in one direction as in the other; so that / the twisting of the axis in the one direction is exactly compensated by the twisting in the opposite direction. We can further see how the tendency is given to make the later formed coils opposite to those, whether turned to the right or to the left, which are first made. Take a piece of string, and let it hang down with the lower end fixed to the floor; then wind the upper end (holding the string quite loosely) spirally round a perpendicular pencil, and this will twist the lower part of the string; and after it has been sufficiently twisted, it will be seen to curve itself into an open spire, with the curves running in an opposite direction to those round the pencil, and consequently with a straight piece of string between the opposed spires. In short, we have given to the string the regular spiral arrangement of a tendril caught at both ends. The spiral contraction generally begins at the extremity which has clasped a support; and these first-formed spires give a twist to the axis of the tendril, which necessarily inclines the basal part into an opposite spiral curvature. I cannot resist giving one other illustration, though superfluous: when a haberdasher winds up ribbon for a customer, he does not wind it into a single coil; for, if he did, the ribbon would twist itself as many times as there were coils; but he winds it into a figure of eight on his thumb and little finger, so that he alternately takes turns in opposite directions, and thus the ribbon is not twisted. So it is with tendrils, with this sole difference, that they take several consecutive turns in one direction and then the same number in an opposite / direction; but in both cases the self-twisting is avoided.

Summary on the nature and action of tendrils

With the majority of tendril-bearing plants the young internodes revolve in more or less broad ellipses, like those made by twining plants; but the figures described, when carefully traced, generally form irregular ellipsoidal spires. The rate of revolution varies from one to five hours in different species, and consequently is in some cases more rapid than with any twining plant, and is never so slow as with those many twiners which take more than five hours for each revolution. The direction is variable even in the same individual plant. In Passiflora, the internodes of only one species have the power of revolving. The vine is the weakest revolver observed by me, apparently exhibiting only a trace of a former power. In the Eccremocarpus the movement is interrupted by many long pauses. Very few tendril-bearing plants can spirally twine up an upright stick. Although the power of twining has generally been lost, either from the stiffness or shortness of the internodes, from the size of the leaves, or from some other unknown cause, the revolving movement of the stem serves to bring the tendrils into contact with surrounding objects.

The tendrils themselves also spontaneously revolve. The movement begins whilst the tendril is young, and is at first slow. The mature tendrils of *Bignonia littoralis* move much slower than the internodes. Generally / the internodes and tendrils revolve together at the same rate; in Cissus, Cobaea, and most Passiflorae, the tendrils alone revolve; in other cases, as with *Lathyrus aphaca*, only the internodes move, carrying with them the motionless tendrils; and, lastly (and this is the fourth possible case), neither internodes nor tendrils spontaneously revolve, as with *Lathyrus grandifloru* and Ampelopsis. In most Bignonias, Eccremocarpus, Mutisia, and the Fumariaceae, the internodes, petioles and tendrils all move harmoniously together. In every case the conditions of life must be favourable in order that the different parts should act in a perfect manner.

Tendrils revolve by the curvature of their whole length, excepting the sensitive extremity and the base, which parts do not move, or move but little. The movement is of the same nature as that of the revolving internodes, and, from the observations of Sachs and H. de Vries, no doubt is due to the same cause, namely, the rapid growth of a longitudinal band, which travels round the tendril and successively bows each part to the opposite side. Hence, if a line be painted along that surface which happens at the time to be convex, the line becomes first lateral, then concave, then lateral, and ultimately again convex. This experiment can be tried only on the thicker tendrils, which are not affected by a thin crust of dried paint. The extremities are often slightly curved or hooked, and the curvature of this part is never reversed; in this respect they differ from the extremities / of twining shoots, which not only reverse their curvature, or at least become periodically straight, but curve themselves in a greater degree than the lower part. In most other respects a tendril acts as if it were one of several revolving internodes, which all move together by successively bending to each point of the compass. There is, however, in many cases this unimportant difference, that the curving tendril is separated from the curving internode by a rigid petiole. With most tendril bearers the summit of the stem or shoot projects above the point from which the tendril arises; and it is generally bent to one side, so as to be out of the way of the revolutions swept by the tendril. In those plants in which the terminal shoot is not sufficiently out of the way, as we have seen with Echinocystis, as soon as the tendril comes in its revolving course to this point, it stiffens and straightens itself, and thus rising vertically up passes over the obstacle in an admirable manner.

All tendrils are sensitive, but in various degrees, to contact with an object, and curve towards the touched side. With several plants a single touch, so slight as only just to move the highly flexible tendril, is enough to induce curvature. *Passiflora gracilis* possesses the most sensitive tendrils which I have observed: a bit of platina wire ⅟₅₀ of a grain (1·23 mg) in weight, gently placed on the concave point, caused a tendril to become hooked, as did a loop of soft, thin cotton thread weighing ⅟₃₂nd of a grain (2·02 mg). With the tendrils of several other plants, loops weighing ⅟₁₆th of / a grain (4·05 mg) sufficed. The point of a tendril of *Passiflora gracilis* began to move distinctly in 25 seconds after a touch, and in many cases after 30 seconds. Asa Gray also saw movements in the tendrils of the cucurbitaceous genus, Sicyos, in 30 seconds. The tendrils of some other plants, when lightly rubbed, moved in a few minutes; with Dicentra in half-an-hour; with Smilax in an hour and a quarter or half; and with Ampelopsis still more slowly. The curling movement consequent on a single touch continues to increase for a considerable time, then ceases; after a few hours the tendril uncurls itself, and is again ready to act. When the tendrils of several kinds of plants were caused to bend by extremely light weights suspended on them, they seemed to grow accustomed to so slight a stimulus, and straightened themselves, as if the loops had been removed. It makes no difference what sort of object a tendril touches, with the remarkable exception of other tendrils and drops of water, as was observed with the extremely sensitive tendrils of *Passiflora gracilis* and of the Echinocystis. I have, however, seen tendrils of the Bryony which had temporarily caught other tendrils, and often in the case of the vine.

Tendrils of which the extremities are permanently and slightly curved, are sensitive only on the concave surface; other tendrils, such as those of the Cobaea (though furnished with horny hooks directed to one side) and those of *Cissus discolor*, are sensitive on all sides. Hence the tendrils of this latter plant, when stimulated / by a touch of equal force on opposite sides, did not bend. The inferior and lateral surfaces of the tendrils of Mutisia are sensitive, but not the upper surface. With branched tendrils, the several branches act alike; but in the Hanburya the lateral spur-like branch does not acquire (for excellent reasons which have been explained) its sensitiveness nearly so soon as the main branch. With most tendrils the lower or basal part is either not at all sensitive, or sensitive only to prolonged contact. We thus see that the sensitiveness of tendrils is a special and localized capacity. It is quite independent of the power of spontaneously revolving; for the curling of the terminal portion from a touch does not in the least interrupt the former movement. In *Bignonia unguis* and its close allies, the petioles of the leaves, as well as the tendrils, are sensitive to a touch.

Twining plants when they come into contact with a stick, curl round it invariably in the direction of their revolving movement; but tendrils curl indifferently to either side, in accordance with the position of the stick and the side which is first touched. The clasping movement of the extremity is apparently not steady, but undulatory or vermicular in its nature, as may be inferred from the curious manner in which the tendrils of the Echinocystis slowly crawled round a smooth stick.

As with a few exceptions tendrils spontaneously revolve, it may be asked, why have they been endowed with sensitiveness? why, when they come into contact / with a stick, do they not, like twining plants, spirally wind round it? One reason may be that they are in most cases so flexible and thin, that when brought into contact with any object, they would almost certainly yield and be dragged onwards by the revolving movement. Moreover, the sensitive extremities have no revolving power as far as I have observed, and could not by this means curl round a support. With twining plants, on the other hand, the extremity spontaneously bends more than any other part; and this is of high importance for the ascent of the plant, as may be seen on a windy day. It is, however, possible that the slow movement of the basal and stiffer parts of certain tendrils, which wind round sticks placed in their path, may be analogous to that of twining plants. But I hardly attended sufficiently to this point, and it would have been difficult to distinguish between a movement due to extremely dull irritability, from the arrestment of the lower part, whilst the upper part continued to move onwards.

Tendrils which are only three-fourths grown, and perhaps even at an earlier age, but not whilst extremely young, have the power of revolving and of grasping any object which they touch. These two capacities are generally acquired at about the same period, and both fail when the tendril is full grown. But in Cobaea and *Passiflora punctata* the tendrils begin to revolve in a useless manner, before they have become sensitive. In the Echinocystis they retain their sensitiveness for some time after they have ceased to / revolve and after they have sunk downwards; in this position, even if they were able to seize an object, such power would be of no service in supporting the stem. It is a rare circumstance thus to detect any superfluity or imperfection in the action of tendrils – organs which are so excellently adapted for the functions which they have to perform; but we see that they are not always perfect, and it would be rash to assume that any existing tendril has reached the utmost limit of perfection.

Some tendrils have their revolving motion accelerated or retarded, in moving to or from the light; others, as with the pea, seem indifferent to its action; others move steadily from the light to the dark, and this aids them in an important manner in finding a support. For instance, the tendrils of *Bignonia capreolata* bend from the light to the dark as truly as a wind-vane from the wind. In the Eccremocarpus the extremities alone twist and turn about so as to bring their finer branches and hooks into close contact with any dark surface, or into crevices and holes.

A short time after a tendril has caught a support, it contracts with some rare exceptions into a spire; but the manner of contraction and the several important advantages thus gained have been discussed so lately, that nothing need here be repeated on the subject. Tendrils soon after catching a support grow much stronger and thicker, and sometimes more durable to a wonderful degree; and this shows how much their internal tissues must be changed. Occasionally it is / the part which is wound round a support which chiefly becomes thicker and stronger; I have seen, for instance, this part of a tendril of *Bignonia aequinoctialis* twice as thick and rigid as the free basal part. Tendrils which have caught nothing soon shrink and wither; but in some species of Bignonia they disarticulate and fall off like leaves in autumn.

Anyone who had not closely observed tendrils of many kinds would probably infer that their action was uniform. This is the case with the simpler kinds, which simply curl round an object of moderate thickness, whatever its nature may be.[8] But the genus Bignonia shows us what diversity of action there may be between the tendrils of closely allied species. In all the nine species observed by me, the young internodes revolve vigorously; the tendrils also revolve, but in some of the species in a very feeble manner; and lastly the petioles of nearly all revolve, though with unequal power. The petioles of three of the species, and the tendrils of all are sensitive to contact. In the first-described species, the tendrils resemble in shape a bird's foot, and they are of no service to the stem in spirally ascending a thin upright stick, but they can seize firm hold of a twig or branch. When / the stem twines round a somewhat thick stick, a slight degree of sensitiveness possessed by the petioles is brought into play, and the whole leaf together with the tendril winds round it. In *B. unguis* the petioles are more sensitive, and have greater power of movement than those of the last species; they are able, together with the tendrils, to wind inextricably round a thin upright stick; but the stem does not twine so well. *B. Tweedyana* has similar powers, but in addition, emits aerial roots which adhere to the wood. In *B. venusta* the tendrils are converted into elongated three-pronged grapnels, which move spontaneously in a conspicuous manner; the petioles, however, have lost their sensitiveness. The stem of this species can twine round an upright stick, and is aided in its ascent by the tendrils seizing the stick alternately some way above and then contracting spirally. In *B. littoralis* the tendrils, petioles, and internodes, all revolve spontaneously. The stem, however, cannot twine, but ascends an upright stick by seizing it above with both tendrils together, which then contract into a spire. The tips of these tendrils become developed into adhesive discs. *B. speciosa* possesses similar powers of movement as the last species, but it cannot twine round a stick, though it can ascend by clasping the stick horizontally with one or both of its unbranched tendrils. These tendrils continually insert their pointed ends into minute crevices or holes, but as they are always withdrawn by the subsequent spiral contraction, the habit seems to us in our ignorance useless. Lastly, / the stem of *B. capreolata* twines imperfectly; the much-branched tendrils revolve in a capricious manner, and bend from the light to the dark; their hooked extremities, even whilst immature, crawl into crevices,

---

[8] Sachs, however (*Text-Book of Botany*, Eng. Translation, 1875, p. 280), has shown that which I overlooked, namely, that the tendrils of different species are adapted to clasp supports of different thicknesses. He further shows that after a tendril has clasped a support it subsequently tightens its hold.

and, when mature, seize any thin projecting point; in either case they develop adhesive discs, and these have the power of enveloping the finest fibres.

In the allied Eccremocarpus the internodes, petioles, and much-branched tendrils all spontaneously revolve together. The tendrils do not as a whole turn from the light; but their bluntly hooked extremities arrange themselves neatly on any surface with which they come into contact, apparently so as to avoid the light. They act best when each branch seizes a few thin stems, like the culms of a grass, which they afterwards draw together into a solid bundle by the spiral contraction of all the branches. In Cobaea the finely branched tendrils alone revolve; the branches terminate in sharp, hard, double, little hooks, with both points directed to the same side; and these turn by well-adapted movements to any object with which they come into contact. The tips of the branches also crawl into dark crevices or holes. The tendrils and internodes of Ampelopsis have little or no power of revolving; the tendrils are but little sensitive to contact; their hooked extremities cannot seize thin objects; they will not even clasp a stick, unless in extreme need of a support; but they turn from the light to the dark, and, spreading out their branches in contact with any nearly flat surface, develop discs. / These adhere by the secretion of some cement to a wall, or even to a polished surface; and this is more than the discs of the *Bignonia capreolata* can effect.

The rapid development of these adherent discs is one of the most remarkable peculiarities possessed by any tendril. We have seen that such discs are formed by two species of Bignonia, by Ampelopsis, and, according to Naudin,[9] by the cucurbitaceous genus *Peponopsis adhaerens*. In Anguria the lower surface of the tendril, after it has wound round a stick, forms a coarsely cellular layer, which closely fits the wood, but is not adherent; whilst in Hanburya a similar layer is adherent. The growth of these cellular outgrowths depends (except in the case of the Haplolophium and of one species of Ampelopsis), on the stimulus from contact. It is a singular fact that three families, so widely distinct as the Bignoniaceae, Vitaceae, and Cucurbitaceae, should possess species with tendrils having this remarkable power.

Sachs attributes all the movements of tendrils to rapid growth on the side opposite to that which becomes concave. These movements consist

[9] *Annales des Sc. Nat. Bot.*, 4th series, vol. xii, p. 89.

of revolving nutation, the bending to and from the light, and in opposition to gravity, those caused by a touch, and spiral contraction. It is rash to differ from so great an authority, but I cannot believe that one at least of / these movements – curvature from a touch – is thus caused.[10] In the first place it may be remarked that the movement of nutation differs from that due to a touch, in so far that in some cases the two powers are acquired by the same tendril at different periods of growth; and the sensitive part of the tendril does not seem capable of nutation. One of my chief reasons for doubting whether the curvature from a touch is the result of growth, is the extraordinary rapidity of the movement. I have seen the extremity of a tendril of *Passiflora gracilis*, after being touched, distinctly bent in 25 seconds, and often in 30 seconds; and so it is with the thicker tendril of Sicyos. It appears hardly credible that their outer surfaces could have actually grown in length, which implies a permanent modification of structure, in so short a time. The growth, moreover, on this view must be considerable, for if the touch has been at all rough the extremity is coiled in two or three minutes into a spire of several turns.

When the extreme tip of the tendril of Echinocystis caught hold of a smooth stick, it coiled itself in a few hours (as described at p. 85) twice or thrice round / the stick, apparently by an undulatory movement. At first I attributed this movement to the growth of the outside; black marks were therefore made, and the interspaces measured, but I could not thus detect any increase in length. Hence it seems probable in this case and in others, that the curvature of the tendril from a touch depends on the contraction of the cells along the concave side. Sachs himself admits[11] that 'if the growth which takes place in the entire tendril at the time of contact with a support is small, a considerable acceleration occurs on the convex surface, but in general there is no elongation on the concave surface, or there may even be a *contraction*; in the case of a tendril of Cucurbita this contraction amounted to nearly one-third of the original length'. In a subsequent passage Sachs seems to feel some difficulty in accounting

---

[10] It occurred to me that the movement of nutation and that from a touch might be differently affected by anaesthetics, in the same manner as Paul Bert has shown to be the case with the sleep movements of Mimosa and those from a touch. I tried the common pea and *Passiflora gracilis*, but I succeeded only in observing that both movements were unaffected by exposure for 1½ hrs to a rather large dose of sulphuric ether. In this respect they present a wonderful contrast with Drosera, owing no doubt to the presence of absorbent glands in the latter plant.

[11] *Text-Book of Botany*, 1875, p. 779.

for this kind of contraction. It must not however be supposed from the foregoing remarks that I entertain any doubt, after reading De Vries' observations, about the outer and stretched surfaces of attached tendrils afterwards increasing in length by growth. Such increase seems to me quite compatible with the first movement being independent of growth. Why a delicate touch should cause one side of a tendril to contract we know as little as why, on the view held by Sachs, it should lead to extraordinarily rapid growth of the opposite side. The chief or sole reason for the belief that the curvature of / a tendril when touched is due to rapid growth, seems to be that tendrils lose their sensitiveness and power of movement after they have grown to their full length; but this fact is intelligible, if we bear in mind that all the functions of a tendril are adapted to drag up the terminal growing shoot towards the light. Of what use would it be, if an old and full-grown tendril, arising from the lower part of a shoot, were to retain its power of clasping a support? This would be of no use; and we have seen with tendrils so many instances of close adaptation and of the economy of means, that we may feel assured that they would acquire irritability and the power of clasping a support at the proper age – namely, youth – and would not uselessly retain such power beyond the proper age. /

CHAPTER V

HOOK AND ROOT-CLIMBERS
CONCLUDING REMARKS

Plants climbing by the aid of hooks, or merely scrambling over other plants – Root-climbers, adhesive matter secreted by the rootlets – General conclusions with respect to climbing plants, and the stages of their development.

Hook-climbers. In my introductory remarks, I stated that, besides the two first great classes of climbing plants, namely, those which twine round a support, and those endowed with irritability enabling them to seize hold of objects by means of their petioles or tendrils, there are two other classes, hook-climbers and root-climbers. Many plants, moreover, as Fritz Müller has remarked,[1] climb or scramble up thickets in a still more simple fashion, without any special aid, excepting that their leading shoots are generally long and flexible. It may, however, be suspected from what follows, that these shoots in some cases tend to avoid the light. The few hook-climbers which I have observed, namely, *Galium aparine, Rubus australis*, and some climbing / roses, exhibit no spontaneous revolving movement. If they had possessed this power, and had been capable of twining, they would have been placed in the class of twiners; for some twiners are furnished with spines or hooks, which aid them in their ascent. For instance, the hop, which is a twiner, has reflexed hooks as large as those of the Galium; some other twiners have stiff reflxed hairs; and Dipladenia has a circle of blunt spines at the bases of its leaves. I have seen only one tendril-bearing plant, namely, *Smilax aspera*, which is

---

[1] *Journal of Linn. Soc.*, vol. ix, p. 348. Professor G. Jaeger has well remarked (*In Sachen Darwin's, insbesondere contra Wigand*, 1874, p. 106) that it is highly characteristic of climbing plants to produce thin, elongated, and flexible stems. He further remarks that plants growing beneath other and taller species or trees, are naturally those which would be developed into climbers; and such plants, from stretching towards the light, and from not being much agitated by the wind, tend to produce long, thin, and flexible shoots.

furnished with reflexed spines; but this is the case with several branch-climbers in South Brazil and Ceylon; and their branches graduate into true tendrils. Some few plants apparently depend solely on their hooks for climbing, and yet do so efficiently, as certain palms in the New and Old Worlds. Even some climbing roses will ascend the walls of a tall house, if covered with a trellis. How this is effected I know not; for the young shoots of one such rose, when placed in a pot in a window, bent irregularly towards the light during the day and from the light during the night, like the shoots of any common plant; so that it is not easy to understand how they could have got under a trellis to the wall.[2] /

Root-climbers. A good many plants come under this class, and are excellent climbers. One of the most remarkable is the *Marcygravia umbellata*, the stem of which in the tropical forests of South America, as I hear from Mr Spruce, grows in a curiously flattened manner against the trunks of trees; here and there it puts forth claspers (roots), which adhere to the trunk, and, if the latter be slender, completely embrace it. When this plant has climbed to the light, it produces free branches with rounded stems, clad with sharp-pointed leaves, wonderfully different in appearance from those borne by the stem as long as it remains adherent. This surprising difference in the leaves, I have also observed in a plant of *Marcgravia dubia* in my hothouse. Root-climbers, as far as I have seen, namely, the ivy (*Hedera helix*), *Ficus repens*, and *F. barbatus*, have no power of movement, not even from the light to the dark. As previously stated, the *Hoya carnosa* (Asclepiadaceae) is a spiral twiner, and likewise adheres by rootlets even to a flat wall. The tendril-bearing *Bignonia Tweedyana* emits roots, which curve half round and adhere to thin sticks. The *Tecoma radicans* (Bignoniaceae), which is closely allied to many spontaneously revolving species, climbs by rootlets; nevertheless, its young shoots apparently move about more than can be accounted for by the varying action of the light.

I have not closely observed many root-climbers, but can give one curious fact. *Ficus repens* climbs up a wall just like ivy; and when the young rootlets / are made to press lightly on slips of glass, they emit

[2] Professor Asa Gray has explained, as it would appear, this difficulty in his review (*American Journal of Science*, vol. xl, Sept. 1865, p. 282) of the present work. He has observed that the strong summer shoots of the Michigan rose (*Rosa setigera*) are strongly disposed to push into dark crevices and away from the light, so that they would be almost sure to place themselves under a trellis. He adds that the lateral shoots, made on the following spring, emerged from the trellis as they sought the light.

after about a week's interval, as I observed several times, minute drops of clear fluid, not in the least milky like that exuded from a wound. This fluid is slightly viscid, but cannot be drawn out into threads. It has the remarkable property of not soon drying; a drop, about the size of half a pin's head, was slightly spread out on glass, and I scattered on it some minute grains of sand. The glass was left exposed in a drawer during hot and dry weather, and if the fluid had been water, it would certainly have dried in a few minutes; but it remained fluid, closely surrounding each grain of sand, during 128 days: how much longer it would have remained I cannot say. Some other rootlets were left in contact with the glass for about ten days or a fortnight, and the drops of secreted fluid were now rather larger, and so viscid that they could be drawn out into threads. Some other rootlets were left in contact during twenty-three days, and these were firmly cemented to the glass. Hence we may conclude that the rootlets first secrete a slightly viscid fluid, subsequently absorb the watery parts (for we have seen that the fluid will not dry by itself), and ultimately leave a cement. When the rootlets were torn from the glass, atoms of yellowish matter were left on it, which were partly dissolved by a drop of bisulphide of carbon; and this extremely volatile fluid was rendered very much less volatile by what it had dissolved.

As the bisulphide of carbon has a strong power / of softening indurated caoutchouc, I soaked in it during a short time several rootlets of a plant which had grown up a plaistered wall; and I then found many extremely thin threads of transparent, not viscid, excessively elastic matter, precisely like caoutchouc, attached to two sets of rootlets on the same branch. These threads proceeded from the bark of the rootlet at one end, and at the other end were firmly attached to particles of silex or mortar from the wall. There could be no mistake in this observation, as I played with the threads for a long time under the microscope, drawing them out with my dissecting needles, and letting them spring back again. Yet I looked repeatedly at other rootlets similarly treated, and could never again discover these elastic threads. I therefore infer that the branch in question must have been slightly moved from the wall at some critical period, whilst the secretion was in the act of drying, through the absorption of its watery parts. The genus Ficus abounds with caoutchouc, and we may conclude from the facts just given that this substance, at first in solution and ultimately modified into an unelastic cement,[3] is used by the *Ficus repens* to cement its rootlets to any surface which it ascends. Whether other plants, which climb by their rootlets, emit any cement I do not know; but the rootlets of the / ivy, placed against glass, barely adhered to it, yet secreted a little yellowish matter. I may add, that the rootlets of the *Marcgravia dubia* can adhere firmly to smooth painted wood.

Vanilla aromatica emits aerial roots a foot in length, which point straight down to the ground. According to Mohl (p. 49), these crawl into crevices, and when they meet with a thin support, wind round it, as do tendrils. A plant which I kept was young, and did not form long roots; but on placing thin sticks in contact with them, they certainly bent a little to that side, in the course of about a day, and adhered by their rootlets to the wood; but they did not bend quite round the sticks, and afterwards they re-pursued their downward course. It is probable that these slight movements of the roots are due to the quicker growth of the side exposed to the light, in comparison with the other side, and not because the roots are sensitive to contact in the same manner as true tendrils. According to Mohl, the rootlets of certain species of Lycopodium act as tendrils.[4] /

Concluding remarks on climbing plants

Plants become climbers, in order, as it may be presumed, to reach the light, and to expose a large surface of their leaves to its action and to that of the free air. This is effected by climbers with wonderfully little expenditure of organized matter, in comparison with trees, which have to support a load of heavy branches by a massive trunk. Hence, no doubt, it arises that there are so many climbing plants in all quarters

[3] Mr Spiller has recently shown (*Chemical Society*, 16 February, 1865), in a paper on the oxidation of india-rubber or caoutchouc, that this substance, when exposed in a fine state of division to the air, gradually becomes converted into brittle, resinous matter, very similar to shellac.

[4] Fritz Müller informs me that he saw in the forests of South Brazil numerous black strings, from some lines to nearly an inch in diameter, winding spirally round the trunks of gigantic trees. At first sight he thought that they were the stems of twining plants which were thus ascending the trees; but he afterwards found that they were the aerial roots of a Philodendron which grew on the branches above. These roots therefore seem to be true twiners, though they use their powers to descend, instead of to ascend like twining plants. The aerial roots of some other species of Philodendron hang vertically downwards, sometimes for a length of more than fifty feet.

of the world, belonging to so many different orders. These plants have
been arranged under four classes, disregarding those which merely
scramble over bushes without any special aid. Hook-climbers are the
least efficient of all, at least in our temperate countries, and can climb
only in the midst of an entangled mass of vegetation. Root-climbers are
excellently adapted to ascend naked faces of rock or trunks of trees;
when, however, they climb trunks they are compelled to keep much in
the shade; they cannot pass from branch to branch and thus cover the
whole summit of a tree, for their rootlets require long-continued and
close contact with a steady surface in order to adhere. The two great
classes of twiners and of plants with sensitive organs, namely, leaf-
climbers and tendril-bearers taken together, far exceed in number and
in the perfection of their mechanism the climbers of the two first
classes. Those which have the power of spontaneously revolving and of
grasping objects with which they come in contact, easily pass / from
branch to branch, and securely ramble over a wide, sunlit surface.

The divisions containing twining plants, leaf-climbers, and tendril-
bearers graduate to a certain extent into one another, and nearly all
have the same remarkable power of spontaneously revolving. Does this
gradation, it may be asked, indicate that plants belonging to one
subdivision have actually passed during the lapse of ages, or can pass,
from one state to the other? Has, for instance, any tendril-bearing
plant assumed its present structure without having previously existed
as a leaf-climber or a twiner? If we consider leaf-climbers alone, the
idea that they were primordially twiners is forcibly suggested. The
internodes of all, without exception, revolve in exactly the same
manner as twiners; some few can still twine well, and many others in an
imperfect manner. Several leaf-climbing genera are closely allied to
other genera which are simple twiners. It should also be observed, that
the possession of leaves with sensitive petioles, and with the consequent power of clasping an object, would be of comparatively little use
to a plant, unless associated with revolving internodes, by which the
leaves are brought into contact with a support; although no doubt a
scrambling plant would be apt, as Professor Jaeger has remarked, to
rest on other plants by its leaves. On the other hand, revolving
internodes, without any other aid, suffice to give the power of climbing;
so that it seems probable that leaf-climbers were in most cases at first
twiners, and subsequently / became capable of grasping a support; and
this, as we shall presently see, is a great additional advantage.

From analogous reasons, it is probable that all tendril-bearers were primordially twiners, that is, are the descendants of plants having this power and habit. For the internodes of the majority revolve; and, in a few species, the flexible stem still retains the capacity of spirally twining round an upright stick. Tendril-bearers have undergone much more modification than leaf-climbers; hence it is not surprising that their supposed primordial habits of revolving and twining have been more frequently lost or modified than in the case of leaf-climbers. The three great tendril-bearing families in which this loss has occurred in the most marked manner, are the Cucurbitaceae, Passifloraceae, and Vitaceae. In the first, the internodes revolve; but I have heard of no twining form, with the exception (according to Palm, p. 29, 52) of *Momordica balsamina*, and this is only an imperfect twiner. In the two other families I can hear of no twiners; and the internodes rarely have the power of revolving, this power being confined to the tendrils. The internodes, however, of *Passiflora gracilis* have the power in a perfect manner, and those of the common vine in an imperfect degree: so that at least a trace of the supposed primordial habit has been retained by some members of all the larger tendril-bearing groups.

On the view hear given, it may be asked, Why have the species which were aboriginally twiners been converted in so many groups into leaf-climbers or tendril-bearers? / Of what advantage has this been to them? Why did they not remain simple twiners? We can see several reasons. It might be an advantage to a plant to acquire a thicker stem, with short internodes bearing many or large leaves; and such stems are ill fitted for twining. Anyone who will look during windy weather at twining plants will see that they are easily blown from their support; not so with tendril-bearers or leaf-climbers, for they quickly and firmly grasp their support by a much more efficient kind of movement. In those plants which still twine, but at the same time possess tendrils or sensitive petioles, as some species of Bignonia, Clematis, and Tropaeolum, it can readily be observed how incomparably better they grasp an upright stick than do simple twiners. Tendrils, from possessing this power of grasping an object, can be made long and thin; so that little organic matter is expended in their development, and yet they sweep a wide circle in search of a support. Tendril-bearers can, from their first growth, ascend along the outer branches of any neighbouring bush, and they are thus always fully exposed to the light; twiners, on the contrary, are best fitted to ascend bare stems, and generally have to start in the shade. Within tall and dense tropical forests, twining plants would probably succeed better than most kinds of tendril-bearers; but the majority of twiners, at least in our temperate regions, from the nature of their revolving movement, cannot ascend thick trunks, whereas this can be affected by tendril-bearers / if the trunks are branched or bear twigs, and by some species if the bark is rugged.

The advantage gained by climbing is to reach the light and free air with as little expenditure of organic matter as possible; now, with twining plants, the stem is much longer than is absolutely necessary; for instance, I measured the stem of a kidney-bean, which had ascended exactly two feet in height, and it was three feet in length: the stem of a pea, on the other hand, which had ascended to the same height by the aid of its tendrils, was but little longer than the height reached. That this saving of the stem is really an advantage to climbing plants, I infer from the species that still twine but are aided by clasping petioles or tendrils, generally making more open spires than those made by simple twiners. Moreover, the plants thus aided, after taking one or two turns in one direction, generally ascend for a space straight, and then reverse the direction of their spire. By this means they ascend to a considerably greater height, with the same length of stem, than would otherwise have been possible; and they do this with safety, as they secure themselves at intervals by their clasping petioles or tendrils.

We have seen that tendrils consist of various organs in a modified state, namely, leaves, flower-peduncles, branches, and perhaps stipules. With respect to leaves, the evidence of their modification is ample. In young plants of Bignonia the lower leaves often remain quite unchanged, whilst the upper ones have / their terminal leaflets converted into perfect tendrils; in Eccremocarpus I have seen a single lateral branch of a tendril replaced by a perfect leaflet; in *Vicia sativa*, on the other hand, leaflets are sometimes replaced by tendril-branches; and many other such cases could be given. But he who believes in the slow modification of species will not be content simply to ascertain the homological nature of different kinds of tendrils; he will wish to learn, as far as is possible, by what actual steps leaves, flower-peduncles, etc., have had their functions wholly changed, and have come to serve merely as prehensile organs.

In the whole group of leaf-climbers abundant evidence has been given that an organ, still subserving the functions of a leaf, may become sensitive to a touch, and thus grasp an adjoining object. With several leaf-climbers the true leaves spontaneously revolve; and their petioles, after clasping a support grow thicker and stronger. We thus see that leaves may acquire all the leading and characteristic qualities of tendrils, namely, sensitiveness, spontaneous movement, and subsequently increased strength. If their blades or laminae were to abort, they would form true tendrils. And of this process of abortion we can follow every step, until no trace of the original nature of the tendril is left. In *Mutisia clematis*, the tendril, in shape and colour, closely resembles the petiole of one of the ordinary leaves, together with the midribs of the leaflets, but vestiges of the laminae are still occasionally retained. In four genera of the Fumariaceae we can / follow the whole process of transformation. The terminal leaflets of the leaf-climbing *Fumaria officinalis* are not smaller than the other leaflets; those of the leaf-climbing *Adlumia cirrhosa* are greatly reduced; those of *Corydalis claviculata* (a plant which may indifferently be called a leaf-climber or a tendril-bearer) are either reduced to microscopical dimensions or have their blades wholly aborted, so that this plant is actually in a state of transition; and, finally, in the Dicentra the tendrils are perfectly characterized. If, therefore, we could behold at the same time all the prognitors of Dicentra, we should almost certainly see a series like that now exhibited by the above-named three genera. In *Tropaeolum tricolorum* we have another kind of passage; for the leaves which are first formed on the young stems are entirely destitute of laminae, and must be called tendrils, whilst the later formed leaves have well-developed laminae. In all cases the acquirement of sensitiveness by the midribs of the leaves appears to stand in some close relation with the abortion of their laminae or blades.

On the view here given, leaf-climbers were primordially twiners, and tendril-bearers (when formed of modified leaves) were primordially leaf-climbers. The latter, therefore, are intermediate in nature between twiners and tendril-bearers, and ought to be related to both. This is the case: thus the several leaf-climbing species of the Antirrhineae, of Solanum, Cocculus, and Gloriosa, have within the same family and even within the same genus, relatives which are twiners. In the / genus Mikania, there are leaf-climbing and twining species. The leaf-climbing species of Clematis are very closely allied to the tendril-bearing Naravelia. The Fumariaceae include closely allied genera which are leaf-climbers and tendril-bearers. Lastly, a species of Bignonia is at the same time both a leaf-climber and a tendril-bearer; and other closely allied species are twiners.

Tendrils of another kind consist of modified flower-peduncles. In this case we likewise have many interesting transitional states. The common vine (not to mention the Cardiospermum) gives us every possible gradation between a perfectly developed tendril and a flower-peduncle covered with flowers, yet furnished with a branch, forming the flower-tendril. When the latter itself bears a few flowers, as we know sometimes is the case, and still retains the power of clasping a support, we see an early condition of all those tendrils which have been formed by the modification of flower-peduncles.

According to Mohl and others, some tendrils consist of modified branches: I have not observed any such cases, and know nothing of their transitional states, but these have been fully described by Fritz Müller. The genus Lophospermum also shows us how such a transition is possible; for its branches spontaneously revolve and are sensitive to contact. Hence, if the leaves on some of the branches of the Lophospermum were to abort, these branches would be converted into true tendrils. Nor is there anything improbable / in certain branches alone being thus modified, whilst others remained unaltered; for we have seen with certain varieties of Phaseolus, that some of the branches are thin, flexible, and twine, whilst other branches on the same plant are stiff and have no such power.

If we enquire how a petiole, a branch or flower-peduncle first became sensitive to a touch, and acquired the power of bending towards the touched side, we get no certain answer. Nevertheless an observation by Hofmeister[5] well deserves attention, namely, that the shoots and leaves of all plants, whilst young, move after being shaken. Kerner also finds, as we have seen, that the flower-peduncles of a large number of plants, if shaken or gently rubbed bend to this side. And it is young petioles and tendrils, whatever their homological nature may be, which move on being touched. It thus appears that climbing plants have utilized and perfected a widely distributed and incipient capacity, which capacity, as far as we can see, is of no service to ordinary plants. If we further enquire how the stems, petioles, tendrils, and flower-peduncles of climbing plants first acquired their power of spontaneously revolving, or, to speak more accurately, of successively bending to all points of the compass, we are again silenced, or at most can only remark that the power of moving, both spontaneously and from various stimulants, is far more / common with plants, than is

[5] Quoted by Cohn, in his remarkable memoir, 'Contractile Gewebe im Pflanzenreiche', *Abhandl. der Schleisischen Gessell.*, 1861, Heft i, p. 35.

generally supposed to be the case by those who have not attended to the subject. I have given one remarkable instance, namely that of the *Maurandia semperflorens*, the young flower-peduncles of which spontaneously revolve in very small circles, and bend when gently rubbed to the touched side; yet this plant certainly does not profit by these two feebly developed powers. A rigorous examination of other young plants would probably show slight spontaneous movements in their stems, petioles, or peduncles, as well as sensitiveness to a touch.[6] We see at least that the Maurandia might, by a little augmentation of the powers which it already possesses, come first to grasp a support by its flower-peduncles, and then, by the abortion of some of its flowers (as with Vitis or Cardiospermum), acquire perfect tendrils.

There is one other interesting point which deserves notice. We have seen that some tendrils owe their origin to modified leaves, and others to modified flower-peduncles; so that some are foliar and others axial in their nature. It might therefore have been expected that they would have presented some difference in function. This is not the case. On the contrary, they / present the most complete identity in their several characteristic powers. Tendrils of both kinds spontaneously revolve at about the same rate. Both, when touched, bend quickly to the touched side, and afterwards recover themselves and are able to act again. In both the sensitiveness is either confined to one side or extends all round the tendril. Both are either attracted or repelled by the light. The latter property is seen in the foliar tendrils of *Bignonia capreolata* and in the axial tendrils of Ampelopsis. The tips of the tendrils in these two plants become, after contact, enlarged into discs, which are at first adhesive by the secretion of some cement. Tendrils of both kinds, soon after grasping a support, contract spirally; they then increase greatly in thickness and strength. When we add to these several points of identity the fact that the petiole of *Solanum jasminoides*, after it has clasped a support, assumes one of the most characteristic features of the axis, namely, a closed ring of woody vessels, we can hardly avoid asking, whether the

[6] Such slight spontaneous movements, I now find, have been for some time known to occur, for instance with the flower-stems of *Brassica napus* and with the leaves of many plants: Sachs' *Text-Book of Botany*, 1875, pp. 766, 785. Fritz Müller also has shown in relation to our present subject (*Jenaischen Zeitschrift*, Bd. V, Heft 2, p. 133) that the stems, whilst young, of an Alisma and of a Linum are continually performing slight movements to all points of the compass, like those of climbing plants.

difference between foliar and axial organs can be of so fundamental a nature as is generally supposed?[7]

We have attempted to trace some of the stages in the genesis of climbing plants. But, during the endless fluctuations of the conditions of life to which all organic beings have been exposed, it might be expected that some climbing plants would have lost / the habit of climbing. In the cases given of certain South African plants belonging to great twining families, which in their native country never twine, but reassume this habit when cultivated in England, we have a case in point. In the leaf-climbing *Clematis flammula*, and in the tendril-bearing vine, we see no loss in the power of climbing, but only a remnant of the revolving power which is indispensable to all twiners, and is so common as well as so advantageous to most climbers. In *Tecoma radicans*, one of the Bignoniceae, we see a last and doubtful trace of the power of revolving.

With respect to the abortion of tendrils, certain cultivated varieties of *Cucurbita pepo* have, according to Naudin,[8] either quite lost these organs or bear semi-monstrous representatives of them. In my limited experience, I have met with only one apparent instance of their natural suppression, namely, in the common bean. All the other species of Vicia, I believe, bear tendrils; but the bean is stiff enough to support its own stem, and in this species, at the end of the petiole, where, according to analogy, a tendril ought to have existed, a small pointed filament projects about a third of an inch in length, and which is probably the rudiment of a tendril. This may be the more safely inferred, as in young and unhealthy specimens of other tendril-bearing plants similar rudiments may occasionally be observed. In the bean / these filaments are variable in shape, as is so frequently the case with rudimentary organs; they are either cylindrical, or foliaceous, or are deeply furrowed on the upper surface. They have not retained any vestige of the power of revolving. It is a curious fact, that many of these filaments, when foliaceous, have on their lower surfaces, dark-coloured glands like those on the stipules, which excrete a sweet fluid; so that these rudiments have been feebly utilized.

One other analogous case, though hypothetical, is worth giving. Nearly all the species of Lathyrus possesses tendrils; but *L. nissolia* is

---

[7] Mr Herbert Spencer has recently argued (*Principles of Biology*, 1865, p. 37 *et seq.*) with much force that there is no fundamental distinction between the foliar and axial organs of plants.

[8] *Annales des Sc. Nat.*, 4th series, Bot. vol. vi, 1856, p. 31.

destitute of them. This plant has leaves, which must have struck every-one with surprise who has noticed them, for they are quite unlike those of all common papilionaceous plants, and resemble those of a grass. In another species, *L. aphaca*, the tendril, which is not highly developed (for it is unbranched, and has no spontaneous revolving power), replaces the leaves, the latter being replaced in function by large stipules. Now if we suppose the tendrils of *L. aphaca* to become flattened and foliaceous, like the little rudimentary tendrils of the bean, and the large stipules to become at the same time reduced in size, from not being any longer wanted, we should have the exact counterpart of *L. nissolia*, and its curious leaves are at once rendered intelligible to us.

It may be added, as serving to sum up the foregoing views on the origin of tendril-bearing plants, that *L. nissolia* is probably descended from a plant which was / primordially a twiner; this then became a leaf-climber, the leaves being afterwards converted by degrees into tendrils, with the stipules greatly increased in size through the law of compensation.[9] After a time the tendrils lost their branches and became simple; they then lost their revolving power (in which state they would have resembled the tendrils of the existing *L. aphaca*), and afterwards losing their prehensile power and becoming foliaceous would no longer be thus designated. In this last stage (that of the existing *L. nissolia*) the former tendrils would reassume their original function of leaves, and the stipules which were recently much developed being no longer wanted, would decrease in size. If species become modified in the course of ages, as almost all naturalists now admit, we may conclude that *L. nissolia* has passed through a series of changes, in some degree like those here indicated.

The most interesting point in the natural history of climbing plants is the various kinds of movement which they display in manifest relation to their wants. The most different organs – stems, branches, flower-peduncles, petioles, midribs of the leaf and leaflets, and apparently aerial roots – all possess this power.

The first action of a tendril is to place itself in a proper position. For instance, the tendril of Cobaea / first rises vertically up, with its branches divergent and with the terminal hooks turned outwards; the

[9] Moquin-Tandon (*Éléments de Tératologie*, 1841, p. 156) gives the case of a monstrous bean, in which a case of compensation of this nature was suddenly effected; for the leaves completely disappeared and the stipules grew to an enormous size.

young shoot at the extremity of the stem is at the same time bent to one side, so as to be out of the way. The young leaves of Clematis, on the other hand, prepare for action by temporarily curving themselves downwards, so as to serve as grapnels.

Secondly, if a twining plant or a tendril gets by any accident into an inclined position, it soon bends upwards, though secluded from the light. The guiding stimulus no doubt is the attraction of gravity, as Andrew Knight showed to be the case with germinating plants. If a shoot of any ordinary plant be placed in an inclined position in a glass of water in the dark, the extremity will, in a few hours, bend upwards; and if the position of the shoot be then reversed, the downward-bent shoot reverses its curvature; but if the stolon of a strawberry, which has no tendency to grow upwards, be thus treated, it will curve downwards in the direction of, instead of in opposition to, the force of gravity. As with the strawberry, so it is generally with the twining shoots of the *Hibbertia dentata*, which climbs laterally from bush to bush; for these shoots, if placed in a position inclined downwards, show little and sometimes no tendency to curve upwards.

Thirdly, climbing plants, like other plants, bend towards the light by a movement closely analogous to the incurvation which causes them to revolve, so that their revolving movement is often accelerated or retarded / in travelling to or from the light. On the other hand, in a few instances tendrils bend towards the dark.

Fourthly, we have the spontaneous revolving movement which is independent of any outward stimulus, but is contingent on the youth of the part, and on vigorous health; and this again of course depends on a proper temperature and other favourable conditions of life.

Fifthly, tendrils, whatever their homological nature may be, and the petioles or tips of the leaves of leaf-climbers, and apparently certain roots, all have the power of movement when touched, and bend quickly towards the touched side. Extremely slight pressure often suffices. If the pressure be not permanent, the part in question straightens itself and is again ready to bend on being touched.

Sixthly, and lastly, tendrils, soon after clasping a support, but not after a mere temporary curvature, contract spirally. If they have not come into contact with any object, they ultimately contract spirally, after ceasing to revolve; but in this case the movement is useless, and occurs only after a considerable lapse of time.

With respect to the means by which these various movements are effected, there can be little doubt from the researches of Sachs and

H. de Vries, that they are due to unequal growth; but from the reasons already assigned, I cannot believe that this explanation applies to the rapid movements from a delicate touch. /

Finally, climbing plants are sufficiently numerous to form a conspicuous feature in the vegetable kingdom, more especially in tropical forests. America, which so abounds with arboreal animals, as Mr Bates remarks, likewise abounds according to Mohl and Palm with climbing plants; and of the tendril-bearing plants examined by me, the highest developed kinds are natives of this grand continent, namely, the several species of Bignonia, Eccremocarpus, Cobaea, and *Ampelopsis*. But even in the thickets of our temperate regions the number of climbing species and individuals is considerable, as will be found by counting them. They belong to many and widely different orders. To gain some rude idea of their distribution in the vegetable series, I marked, from the lists given by Mohl and Palm (adding a few myself, and a competent botanist, no doubt, could have added many more), all those families in Lindley's *Vegetable Kingdom* which include twiners, leaf-climbers, or tendril-bearers. Lindley divides Phanerogamic plants into fifty-nine Alliances; of these, no less than thirty-five include climbing plants of the above kinds, hook and root-climbers being excluded. To these a few cryptogamic plants must be added. When we reflect on the wide separation of these plants in the series, and when we know that in some of the largest, well-defined orders, such as the Compositae, Rubiaceae, Scrophulariaceae, Liliaceae, etc., species in only two or three genera have the power of climbing, the conclusion is forced on our minds that the capacity of revolving, on which most climbers depend, is inherent, / though undeveloped, in almost every plant in the vegetable kingdom.

It has often been vaguely asserted that plants are distinguished from animals by not having the power of movement. It should rather be said that plants acquire and display this power only when it is of some advantage to them; this being of comparatively rare occurrence, as they are affixed to the ground, and food is brought to them by the air and rain. We see how high in the scale of organization a plant may rise, when we look at one of the more perfect tendril-bearers. It first places its tendrils ready for action, as a polypus places its tentacula. If the tendril be displaced, it is acted on by the force of gravity and rights itself. It is acted on by the light, and bends towards or from it, or disregards it, whichever may be most advantageous. During several days the tendrils or internodes, or both, spontaneously revolve with a steady motion. The tendril strikes some object, and quickly curls round and firmly grasps it. In the course of some hours it contracts into a spire, dragging up the stem, and forming an excellent spring. All movements now cease. By growth the tissues soon become wonderfully strong and durable. The tendril has done its work, and has done it in an admirable manner. /

INDEX

Printed and bound by CPI Group (UK) Ltd, Croydon, CR0 4YY

23/10/2024

01777667-0020